COUNTERING DISPOSSESSION, RECLAIMING LAND

COUNTERING DISPOSSESSION,

RECLAIMING LAND

A SOCIAL MOVEMENT ETHNOGRAPHY

David E. Gilbert

UNIVERSITY OF CALIFORNIA PRESS

University of California Press
Oakland, California

© 2024 by David Gilbert

Library of Congress Cataloging-in-Publication Data

Names: Gilbert, David E., 1982- author.
Title: Countering dispossession, reclaiming land : a social movement
    ethnography / David E. Gilbert.
Description: Oakland, California : University of California Press,
    [2024] | Includes bibliographical references and index.
Identifiers: LCCN 2023041705 (print) | LCCN 2023041706 (ebook) |
    ISBN 9780520397750 (cloth) | ISBN 9780520397767 (paperback) |
    ISBN 9780520397774 (epub)
Subjects: LCSH: Environmental justice—Indonesia—Case studies. |
    Land tenure—Indonesia. | Plantations—Indonesia. | Plantation
    workers—Political activity—Indonesia.
Classification: LCC GE240.I5 .G55 2024 (print) | LCC GE240.I5 (ebook)
    | DDC 304.2/809598—dc23/eng/20231019
LC record available at https://lccn.loc.gov/2023041705
LC ebook record available at https://lccn.loc.gov/2023041706

33 32 31 30 29 28 27 26 25 24
10 9 8 7 6 5 4 3 2 1

For Zarah Maria

A revolution that is based on people exercising their creativity in the midst of devastation is one of the great historical contributions of humankind.

—Grace Lee Boggs, 2014

CONTENTS

DRAMATIS PERSONAE

Agoez	Plantation laborer turned reclaimer; avid sugar palm tapper
Citra	Migrant laborer in the city turned back to the lander
Kuman	Migrant oil palm plantation laborer turned reclaimer
Midwal	Aren volcano woodsman and reclaimer
Nurul	Community Council representative and anarchist organizer
Riza	Reclaiming organizer, village head, and defendant in Dona Company's lawsuit against the reclaiming movement
Rudi	Minangkabau philosopher and forest farmer
Susilo	Reclaiming activist, cinnamon forest farmer, and manager (*kolektor*) of the collective land
Tia	Landowner, reclaimer, and blockader; member of the Flamboyant Smallholder Women's Group cooperative
Wangga	Plantation laborer and migrant turned reclaimer; agroecologist
Wibawa	Reclaiming movement archivist, religious scholar, and bureaucrat
Zed	Reclaiming movement organizer and blockader

PEASANT UNION MEMBERS

Aren	Indonesian Peasant Union national organizer and lawyer, presciently named for the Aren volcano he was born on
Daud	Casiavera mosque manager and Indonesian Peasant Union organizer on the reclaimed land
Lufti	Migrant laborer, Indonesian Peasant Union member, and reclaimer; active tree planter on his wife's family land and the collective land
Malin	Founding organizer of the West Sumatran Peasant Union; reclaimer and dairy cooperative member on the collective land
Mansur	Indonesian Peasant Union member, reclaiming movement fruit trader in Casiavera, and barber
Henry Saragih	Founding member and chair of the Indonesian Peasant Union; coordinator of Via Campesina, international social movement umbrella organization

COLONIZERS

Abril	Plantation manager; village head of Casiavera during the New Order military dictatorship
J. Ballot	Dutch governor of West Sumatra; fired for refusing to uphold the dispossession of West Sumatra's Minangkabau Nation
J. H. Chaping	Dutch forestry official; demarcated state lands across Casiavera and the Aren volcano
Dona Company	Indonesian agribusiness that took control of the land in 1968; operated a cattle ranch and ginger and tobacco plantations into the 1990s
Allen W. Dulles	Central Intelligence Agency director; authorized a covert war in West Sumatra to overthrow the newly established Indonesian Republic in the 1950s
Irfan	New Order Ministry of Forestry official; implemented the Pine Plantation Program in Casiavera
National Land Agency	Arm of the Indonesian state tasked with creating land concession leases for the corporate dispossession and exploitation of the land

Judge Johan Pyper	Judge who signed the land concession lease that authorized the Dutch colonial-era dispossession of Casiavera's collective land in the early 1900s
C. P. C. Steinmetz	Colonial Resident of the West Sumatran highlands; oversaw the forced cultivation of coffee in Casiavera at the turn of the nineteenth century
Mahmud Teuling	First owner of Dona Company, one-time military police officer, and cigarette trader
Monica Jessica Teuling	World Bank accountant and second owner of the Dona Company; daughter of Mahmud Teuling
W. H. Samuel Company	German agribusiness that held the first land lease on Casiavera's collective land starting in 1905
Yono	New Order Forest police officer

Land Back

In late 2012, a delegation of hundreds of activists from more than twenty-five nations gathered high up on the Aren volcano in Sumatra's Bukit Barisan mountains. The Indonesian Peasant Union (Serikat Petani Indonesia), one of Indonesia's largest agrarian justice organizations, worked with The Peasants' Way (La Via Campesina), an international activist organization, to bring them together.

Indigenous peoples joined agricultural laborers, smallholder farmers, and politicians from as near as South Sumatra and as far as Senegal and Brazil. A few campaigners, artists, and scholars, me among them, joined as well. The delegation traveled to Casiavera, a village on the volcano, to discuss their shared struggles for livelihood and well-being as states and corporations across the world remain intent on evicting rural, Indigenous, and marginalized peoples from their homes and dispossessing them of their forests, agricultural lands, and waterways.

Organizers with the Indonesian Peasant Union chose to host everyone in Casiavera because it is the site of a remarkable land back movement of smallholder family farmers and plantation laborers. Starting first in the late 1990s, a group of agriculturalists from Casiavera worked together to occupy the Dona Company cattle ranch and plantation above their homes. Ever since, Casiavera's reclaimers have worked to expel the agribusiness company, challenge the government's power to control their land, and make a new life working it.

The opening speaker at the gathering, a co-founder of the Indonesian Peasant Union and coordinator of Via Campesina, celebrated Casiavera's reclaiming movement and urged the visiting delegation "to take inspiration from these reclaimers for the continuation of our movements into the twenty-first-century."

After three days of visits with Casiavera's movement members and long conversations among themselves, the visiting Via Campesina delegation released their Bukit Tinggi Declaration. The declaration outlined their shared vision of modern global smallholder movements, which work to "defend land and territory, erase poverty, and honor the earth."[1] That the delegation chose to release this vision of agrarian justice from Casiavera spoke to the relevance of this reclaiming movement for other struggles for livelihood and autonomy across the Earth's many landscapes of extraction and exploitation.

Moved by my brief visit to Casiavera in 2012, I returned to Aren in 2015 to live and work there for the better part of a year.[2] Malin, a smallholder and longtime peasant union member whom I had met at the Via Campesina gathering, offered to introduce me to life on the reclaimed land. Directly behind his family's home in the middle of the village was a slight rise, where I looked out on a tall forest growing across a miles-long hollow of the volcano. Above this gently sloping hollow, the volcano's slopes steepened again to become the deeper green of the cloud forests on the mountain's highest reaches. At the edge of town just before the start of these forests, a peculiar gate was visible. Across the top of the gate were the words "Collective Land" (*Tanah Ulayat*) written in big, black block letters. "Up until the late 1990s the gate said Dona Company," Malin told me, pointing at the sign.

After passing through the gate into what Casiavera's reclaimers now call their "collective land" (*tanah ulayat*), a twenty-minute walk led us to Malin's family plot, where with his wife he tended a small plot of cacao, cinnamon, clove, avocado, sugar palm, and mahogany trees.[3] Malin's family plot was one of more than two hundred similar smallholder plots on the land. Interspersed between them were vegetable gardens and patches of forest. Nearby was a fifteen-cow dairy shed, where a handful of younger women and men were sitting on buckets, milking their cows. A few simple barns belonging to agricultural cooperatives marked the upper reaches of the land.

It was a bustling landscape, alive with thousands of trees that produce valuable food, fiber, and timber. All of this, Malin told me, began in the 1990s with the work of smallholders, a peasant union, and a few small cooperatives. Under the cool shade of the broken canopy, the sounds of groups

FIGURE 1. The gate leading to Casiavera's reclaimed land. The letters written across the gate now read *Tanah Ulayat*, or Collective Land. Up until the late 1990s the gate read Dona Company.

of reclaimers working their plots traveled out from other parts of the land, along with birds moving about higher up in the trees. The spicy-sweet smell of cinnamon bark hung in the air. Scattered across the land were a few still uncultivated plots, a reminder of how Casiavera's newfound smallholders planted their forests in what was once a plantation without any trees at all.

Countering Dispossession, Reclaiming Land is an ethnography of the social movement that unfolded in Casiavera. I focus on reclaiming as a way to counter dispossession: as a mobilization away from state and corporate exploitation of the land and toward a small-scale, cooperative, and collective life. I ask how Casiavera's reclaiming movement emerged to counter a century-long history of multiple dispossessions of smallholders from the land. Along the way, I inquire about how this movement created new social and ecological relations, remaking the community into a smallholder economy and the land into a smallholder landscape.

Eventually, in Casiavera hundreds of onetime plantation laborers, landless workers, and smallholders took collective control of the land and cultivated a forest ecology on it that supported their own, albeit imperfect, emancipation. Casiavera's transformation is an accomplishment that deserves

FIGURE 2. A new agricultural forest grows on the collective land. Planted sugar palm, siri, banana, clove, chocolate, and avocado grow. Bordering this planted forest is a still-unplanted plot with grass for cattle (foreground).

celebration, even while reclaimers faced challenges and outright failure. Nearly half of the families who tried to reclaim the land failed to make their plots productive. Other reclaimers were never able to get access to a plot, even though they wanted one. More difficult still was that even after they gained control of the land, many of Casiavera's reclaimers remained in relations of debt, contract farming, and wage labor—the very relations of work that reclaimers sought to avoid when they set out to become smallholders. Still, over the last two decades hundreds of women and men in equal measure have gained access to a one-quarter hectare plot on the collective land, worked to cultivate it, founded cooperatives, and joined up with a peasant union to create a new political agroecology across the land.

FROM DISPOSSESSION TO RECLAIMING

When I first met Malin, he told me only that in the late 1990s workers in Casiavera reclaimed the land from "investors." Over the months and years that followed, Malin led me to understand how the struggle for this land involved a much fuller history that includes a series of radical transformations. In

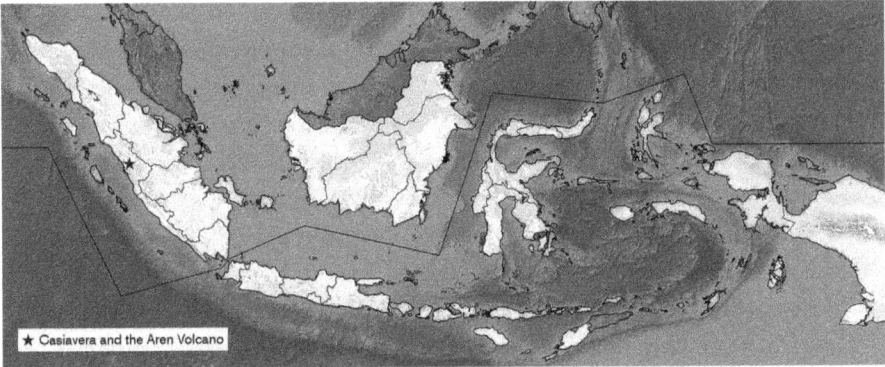

MAP 1. The Indonesian archipelago and Casiavera on the Aren volcano.

the mid-1800s a new form of Dutch colonialism brought a draconian forced coffee cultivation scheme to the volcano. Colonial overseers used violence to compel Casiavera's residents to clear the cloud forests above their homes and plant coffee trees. Dutch-forced coffee cultivation would eventually end in the face of entrenched Indigenous resistance fifty years later.

Following the failure of forced cultivation, in 1905 the Dutch Council of Justice in Padang for the first time signed over the legal right to control the land itself to a European agribusiness, W. H. Samuel. The company took control of the land, logged the forest, and established a cattle ranch.

Generations of company control of the land were broken during World War II when the brutal Japanese occupation of the archipelago ruptured more than three centuries of Dutch colonialism in Sumatra. At the end of the war, an Indigenous revolutionary movement was able to defeat the Dutch army's attempts to regain control of the archipelago, establishing the post-colonial Indonesian Republic. The birth of the republic was a time of smallholder liberation from the colonial plantations and factories. On the Aren volcano Casiavera's smallholders were able to cultivate their lands, previously lost to colonial dispossession, in peace for some fifteen short years.

Dispossession returned in full force in 1968, after the fall of the republic and the start of Indonesia's murderous New Order military dictatorship, when the national Agrarian Directorate leased the land to the Dona Company and its owner, Mahmud Teuling, a retired military police officer. For a full three decades, company staff ruled the land and oversaw a cattle ranch and tobacco plantation.

The full weight of the New Order, one of the twentieth century's most violent military dictatorships, backed Teuling's control of the land and the laborers who worked it. Forged out of a genocide that targeted members of leftist political parties and workers' organizations, the New Order perfected the use of force to dispossess smallholders. Teuling's ranch and plantation was a merciless place. The forests were cleared. Teuling became rich. Two decades later, in the early 1990s, Teuling's daughter took control of the land concession and started a ginger plantation. Erosion from sun and rain and the spraying of herbicides and pesticides damaged the soil. Organic matter and nutrients were lost. People, plants, animals, insects, and microorganisms were poisoned.

Where Casiavera was once a community of free smallholders, under colonial and authoritarian rule the latter became dispossessed laborers. Eventually the logging, cattle ranching, and monocultures on the land ruined it, leaving the soil bare, leeched, compacted, and eroded. Throughout this era, Casiavera's residents remembered that they had cultivated the land above their homes through different periods of this changing history of land control. And when the New Order weakened in the 1990s, hundreds from Casiavera went up and occupied the plantation as part of an archipelago-wide reclaiming movement.

In a series of letters to government officials, Casiavera's reclaimers declared the Dona Company bankrupt and the plantation land ruined, even as the company refused to abandon its claims to the land. Reacting to Casiavera's first occupations along the edges of the plantation, Dona ownership and staff sought to continue planting its ginger and tobacco monocultures. They announced plans for a new industrial sorghum plantation on the land. But in the late 1990s hundreds more reclaimers from Casiavera joined the occupation of the land, effectively bringing the plantation to an end.

Early on in Casiavera's reclaiming, when the first families went up to the plantation land to cultivate it as their own, activists and political leaders took note. One of the first peasant unions to form under the New Order, the West Sumatran Peasant Union (Serikat Petani Sumatera Barat), then still operating underground, began recruiting members in Casiavera. The union worked to strengthen the occupation with protest organizing, agroecology training, and legal support. At the same time, movement leaders brought other movement members to Casiavera, framing Casiavera's mobilization as one that could inspire and inform others.

So began two decades of connection between Casiavera's reclaiming and Indonesia's agrarian movements. The West Sumatran Peasant Union would eventually join up with similar regional organizations to create the national Indonesian Peasant Union, a movement organization that has coordinated reclaiming movements for decades across the archipelago.

Casiavera's reclaiming movement was about regaining control of land and territory, what the latest generation of Indigenous activists in the Americas call land back: the anti-colonial (re)creation of Indigenous land control, community, and landscapes. As such, Casiavera's Indigenous Minangkabau-led reclaiming movement joins the efforts of hundreds of thousands of landless agricultural workers and smallholder farmers, Indigenous and not, in Indonesia and beyond who are struggling to reclaim plantations and other sites of state and corporate dispossession of land (e.g., North America's Land Back, the Zapatistas, the Landless Workers' Movement in Brazil, the Kenyan Peasant League, and the Taiwan Farmers Union).[4] These are all movements for agrarian sovereignty, struggles for the right of rural peoples of all kinds to determine for themselves why, when, and where they live and work.

Reclaiming movements seek to create new livelihoods and ecologies that move away from what has been the dominant form of economy in the Sumatran countryside for over a hundred years: the state handover of land to corporations for industrial exploitation. These mines, timber estates, dams, and plantations reordered the landscape seemingly without moral concern.

Today's dispossessions continue the killing of thousands of species of living beings and push people off their land. Over the last fifty years, more than two-thirds of Sumatra's forests have fallen as voracious logging operations and plantations of palm oil and timber spread apace across the island.[5] The agrarian changes have brought about the tragic impending extinction of the Sumatran rhino and orangutan, arson of plantation infrastructures, militarized repression against dissenters, and the murder of activists working toward another way of doing things.

Violent overlords, capitalist financiers, men and women laboring as coolies, ranch hands, planters, pickers, pesticide sprayers, and deforestation: Casiavera's reclaiming movement brought all of these ills of ranching and plantation production to an end. As a way of countering their own dispossession, Casiavera's reclaimers constructed new, agroecological lifeways that centered on diversified work and cooperatives. A smallholder landscape took form as reclaimers took control of the land and cultivated agricultural

FIGURE 3. Forest death for a new oil palm plantation in Sumatra's Bukit Barisan mountains, not far from Casiavera.

forests on it, what people on the volcano call *parak* and Westerners call agroforests.

Casiavera's agroforests have a threefold importance related to nature, global climate cycles, and smallholder economics. Intimate, human-scale relations with commodity-giving plants grown without toxic chemicals were the foundation of reclaimers' agroecological livelihoods. These were reciprocal relations that made the landscape healthier. The agroforests are significant for their ecological diversity: about half that of the uncultivated cloud forest that grows on the highest reaches of the Aren volcano, but far greater than that of the industrial monoculture plantations that now cover hundreds of thousands of square miles of Sumatra. What's more, Casiavera's growing forests made up a watershed, completing Sumatra's complex water cycles in ways that industrial agriculture does not. These cycles allowed the agroforests to take up carbon out of the atmosphere. In contrast, industrial agriculture is the world's second largest emitter of carbon dioxide and other climate-changing gasses after the energy sector.[6]

Casiavera's forests are perhaps best understood as an expression of reclaimers' specific movement worldviews, which are not reducible to explicitly material or economic concerns. These lively working food forests link unstable

FIGURE 4. The Sumatran artist Wiyono and I collaborated to create these illustrations of how Casiavera's reclaimers transformed the land from cattle ranch and monoculture plantation (top) to smallholder agroecological landscape (bottom) between 1970 and 2015.

matters of political mobilization, household production, and human ecology. They are emergent from reclaimers' unique critiques of industrial capitalist agriculture and participation in direct-action land protest. Reclaimers honed these critiques with their deeply held Indigenous Minangkabau ideas of what they call their matriarchate, a matrilineal form of community organization that places women's rule at the center through practices of collective land control, social deliberation and consensus seeking, and environmental balance upheld through cycles of creation and destruction. Reclaimers in Casiavera acted along moral, ideological, and spiritual dimensions to elaborate nothing less than an agrarian cosmology.

Casiavera's reclaiming attests to the fact that colonial and capitalist dispossession is a flawed enterprise. It cannot expand forever. Yet reclaiming movements also carry import beyond critiques of dispossession. The changes that unfolded across Casiavera, on the Aren volcano, and in other locales of the Bukit Barisan are nothing less than the resurgence of smallholder life. Reclaiming in Casiavera was the way rural workers sought out better livelihoods than the typically difficult, dangerous, and exploitative forms of agrarian work available to them in the surrounding plantations, logging operations, and mines. The many social movements working for decolonization, food sovereignty, land for the tiller, indigenous rights, and the environment are all in need of more dialogue on how to counter dispossession and move toward new lifeways, uncertain as these new topographies of life remain. My aim is to show that reclaimers' experiences on the Aren volcano can provide a point of reference in these discussions.

RECLAIMING! CREATING COUNTER-DISPOSSESSIONS

Critical agrarian studies, a collection of anthropology, sociology, and human geography, provided me with a set of concepts to understand changes in Casiavera. The best of these studies have shown how capitalist relations emerge among agrarian peoples as state planners, financiers, and industry continuously remake landscapes into zones of extraction and exploitation. These changes are the *agrarian question*, which scholars have recast as questions about changes in land, labor, and capital in the countryside for more than a century.[7] One enduring component of these analyses is dispossession, specifically the expropriation of land and its commodification. As it has unfolded across the planet over the last three centuries, *dispossession* is a process of state and corporate territorial acquisition that targets Indigenous and agrarian peoples' lands for enclosure.[8] For both Alexander Chayanov, the early theorist of Russian peasant collectives, and Pierre-Joseph Proudhon, one of the first anarchist scholars, dispossession was another word for the state theft of peasant lands. For Karl Marx, dispossession was the way that the capitalist relation insinuated itself into Indigenous and peasant ways of life.[9]

Dispossessions' harms are manifold. They extend out from the exclusionary enclosures of Indigenous and smallholder lands to include the use of waged labor under varying forms of compulsion and the ecological disasters of extinction and contamination that accompany clear-cutting the forests, mining, and agriculture powered with fossil fuels and agrichemicals. Dispossession is one cause of environmental damage made possible when the state hands over land for large-scale, capitalist exploitation. At the same time, dispossession is also one cause of landlessness and smallholder impoverishment. When they lose access to the resources that once supported their livelihoods, agrarian peoples' communities are fractured.

Three generations of scholarship have established the linkages between dispossession and the destruction of watersheds, the extinction and loss of many forms of life, and human violence and suffering.[10] I have drawn similar conclusions from my visits to and studies of places like Socfin—the first palm oil plantation in Sumatra, established in 1911, a Dutch, Belgian, and British agribusiness—as well as in more recent plantation expansion zones overseen by the modern transnational agribusiness Wilmar, a global corporation incorporated in Singapore, and Cargill, headquartered in Minnesota. The many connected ways dispossession causes all of these ills are complex processes related to the rise of the globalized commodity economy and agrarian differentiation, in which inequalities within and between different social classes in the countryside grow.

The colonial territories of the South took form through multivalent forms of dispossession, including military occupation, annexation, and purchase. State and corporate elites developed tools to steer and profit from the changes. Like all colonial governments, Dutch occupiers of the Indonesian archipelago constantly created new land law to support their regimes of dispossession. Authorities reconceptualized Indonesia's lands many times over as abstract legal entities like forest reserves and plantation land leases. These land control systems enabled the colonial state to hand over land to the early multinational corporations specializing in mining, logging, and plantation agriculture.

State administrators and capitalists have long justified their dispossessions as a means to increase the productivity of the land and grow national economies. Most influential of all proponents of dispossession, Lenin

considered the Communist-led takeover of peasant lands as the only acceptable path toward two socially preferable agricultural achievements: yields
and efficiency. The premise was that state and corporate control of the land
would bring prosperity to the population as a whole by increasing yields
while at the same time forcing the populace to become more urban.[11]

Whatever their apparent differences from Lenin's anti-peasant politics,
many modern-day state planners, agribusiness executives, and agronomists
continue to repeat this legitimization, that agrarian peoples must be dispossessed to make way for the modern development of the countryside. Accordingly, neocolonial and corporate narratives often revolve around the premise
that urbanizing populations are a marker of progress.[12] Agriculture is to
employ ever fewer people who are ever less connected to the land, replacing anachronistic smallholders, who are to move to the city and start again
there.[13] Smallholders are said to be a social group fated to disappear, sentenced to the dustbin of history.[14]

From the earliest periods of capitalist dispossession through today, agrarian scholars have refuted this sorry, unilinear teleology of agrarian change.[15]
Even as Indonesia urbanizes and the percentage of household incomes earned
from agriculture shrinks, the total number of agrarian peoples continues to
increase, including smallholder farmers, landless agricultural laborers, and
fisherfolk. The smallholder—mobile, flexible, and engaged in many kinds of
work while using the land as a home and for agriculture—endures.[16] What's
more, smallholders and the landless alike strive to reclaim and maintain
their position in the face of land grabs, historic and ongoing, that have
sought to force them from the land. Long histories of dispossession have
not erased these peoples. Reclaimers are working to bring the current era of
dispossession to an end.

A subset of Indonesia's smallholders has already moved beyond the impe-
rial and neocolonial boundaries of state and capitalist dispossessions, having
reclaimed industrial logging concessions as well as timber and oil palm plantations. For these people, dispossession at the hands of the government and
industry is not the final endpoint of agrarian change. Smallholder persistence and reemergence (repeasantization) require theorizations not defined
with colonial and capitalist unilinear logics. Reclaiming movements proceed
in fits and starts. They bring reversions to old forms and divergences into
new, unanticipated types. These movements suggest agrarian changes are
"bushy," taking a multiplicity of branches leading to an uncountable number

of contingent forms through time.[17] The interplay of mutually constituted political-economic structures and localized cultural agency influences unending processes of reconfiguration.

Nearly everywhere dispossession and industrial resource exploitation has arrived, there are ruined, damaged places. While the manifold problems of extractivism are known, academic work on the undoing of ruin and the *reclaiming* of lands previously lost to dispossession remains nascent. Heightening the need to understand how to counter agrarian dispossession is the reality that the early twenty-first century is a moment of agrarian redux. The global deregulation of commodity markets has spurred post-colonial states to shift yet again to export-oriented agriculture, logging, and mining, in the process sparking waves of enclosures and dispossession.

As this agrarian redux began in the mid-1990s, the rural sociologist Philip McMichael provided a prescient analysis of the questions facing the current millennium.[18] With a new round of plantations pushing people off the land to destroy unique and already threatened ecologies, the agrarian question became how rural workers can construct a politics and practice of reclaiming. That is, how can smallholders and the landless improve their position vis-à-vis agribusiness and the international finance capital that enables it? Here, the agrarian question shifts from being about changes that stem from state and industry dispossession to a question about what happens after such dispossession.

A profound insight that built upon this scholarship and related organizing was gaining strength across the globally networked Indigenous and agrarian justice movements. In Indonesia, a coalition of smallholder unions across Sumatra, Java, the Eastern Islands, and West Papua joined with the Indonesian Legal Aid Foundation (Yayasan Lembaga Bantuan Hukum Indonesia) to announce the existence of a national "Reclaiming Movement" (Gerakan Reklaim) in 2001. In support of this national mobilization, twenty-six unions and solidarity organizations contributed essays to one of the first Indonesian books on agrarian movements to be published following the end of the New Order, titled *Reclaiming and Peoples' Sovereignty* (*Reklaiming dan Kedaulatan Rakyat*). In it, the writers presented accounts of a collection of non-violent smallholder and Indigenous peoples' reclaiming movements across the archipelago that, in a matter of just two or three years, "grabbed back" (*mengambil lagi*) more than one hundred thousand hectares of land from plantations and logging estates.

The authors wrote about the police arrests and extrajudicial violence they faced from the very same capitalists and enforcers that carried out the dispossession of agrarian peoples during the New Order. They lamented the stigma society poured upon jailed reclaimers and the terrible criminal, paramilitary, and state violence imposed upon them. They also celebrated their belief that for all the repression and criminalization, the reactionary responses to their mobilizations had "not been able to stem the swift, strong flow of demands for justice of smallholders."

In its entirety, *Reclaiming and Peoples' Sovereignty* made clear that for many agrarian peoples, the agrarian question was foremost about political action and how they could work as agile collections of collectives, organizations, and movements capable of redefining economics and politics. Such reclaiming movements require people to reject overly determined ideas of development and move against, away from, and beyond trajectories of dispossession. Reclaiming is counter-dispossession when it involves workers' acknowledgment of the many problems of capitalist industry in the countryside as well as a concerted effort to move beyond the industrial exploitation of land and labor into a modern smallholder life and landscape. In this way, reclaiming movements are ecosocial transformations that challenge the normative practices and values of existing capitalist political economies.

Fundamentally non-violent but not wholly averse to militancy, millions of workers across the world have since organized to reclaim the land. These movements give the countryside its anti-capitalist and anti-imperialist potential. When these workers carry hoes and banners to provocative street protests and land occupations, they put reclaiming at the center of the modern agrarian question.[19] Ensconced in the Lacandona cloud forests, the Zapatista theoretician Subcomandante Marcos wrote lyrically about what many Indigenous peoples, landless, and smallholders hope to create in place of the hacienda, logging concession, and plantation. In response to critiques that reclaiming movements are misguided efforts of repeasantization based on outmoded ideas of "returning to the land," Marcos refuted the idea that reclaimers are attempting to return to a "traditional" political economy:

> We are not trying to return to the past. . . . The struggle is not pointing backwards. In a linear world, where above is considered eternal and below inevitable, the Indian peoples of Mexico are breaking with that line and pointing towards something which is yet to be deciphered, but which is already new and better.[20]

Even Marcos, activist-theoretician par excellence, could not elaborate pre-cisely what the Zapatista movement sought to construct. But Marcos knew linear ideas about agrarian development could not describe what reclaimers wanted to achieve. The political-ecological landscapes reclaimers constructed with land occupation and collectivist agroecology are not old, traditional, nor industrial; they are still undetermined, singularly innovative smallholder spaces.

The Zapatista experience can be held up alongside events in Casiavera to specify a few components of dispossessed peoples' visions of reclaiming. These are groups with deep histories of industrial capitalism that seek to control land lost to corporations. They want the land back so they can use it as a basis of small-scale economic production and dwelling for their families. Reclaimers work to forge many kinds of collectivity, with land occupation, cooperatives, unions, and land ownership. These efforts of economic activity within reclaimed spaces require social organization. Land control and mobi-lization organizations are necessarily constructed. Community and move-ment councils mediate these mobilizations.

Reclaiming involves undoing dispossession. Nevertheless, countering dis-possession is only a starting point in the process of reclaiming. It is what people do with the land once they have control of it that defines their liveli-hood and well-being. In Casiavera, liberation from a century of dispossession included creating a smallholder economy. Along the way, Casiavera's reclaim-ers undid the ecological destruction that had come with their dispossession.

Some disparage reclaiming movements like those unfolding in Chiapas and Aren as an impossibility, as projects of unrealistic and utopian dreams. It is true that only a minority of thinkers join Marcos, the scholars of the Indonesian Legal Aid Foundation, Philip McMichael, and me to see non-industrial agrarian places for what they offer, not what they lack. Yet scholars of modern peasant movements, especially, have done much to show the vibrant importance of agrarian lifeways founded on smallholder economies and ecologies.[21]

The relevance of these modern peasant movements has only grown as modernity increasingly revolves around learning to live with processes of dispossession. In these places, the most pressing questions have become about how to create new lives in the ruined wake of extractive political economies and destroyed landscapes. The magnitude of the many ills of the capitalist exploitation of workers and ecologies demands a reconsideration of how to bring economic activity into balance with social and environmental well-being.

Most promising among concepts of how to do so in the Global North is degrowth, a concept that seeks alternatives to capitalism to decarbonize, detoxify, and attenuate the inequalities of economic production. Rejecting the primacy of unending capitalist economies, degrowth begins from the idea that capitalism as it exists today can't continue forever. Indeed, along current trajectories of capitalist exploitation, it seems not long now before the air and water will be too toxic, the climate too unstable and extreme, and societies too inequal for global social well-being to be a reality.

Building on long-established Marxist critiques of capitalism, the market, and development, degrowth locates the connected economic and ecological problems of society in the violent, unequal, and wasteful commodification of people and nature as resources to be exploited. To overcome these problems, degrowth seeks nothing less than to break apart the already fissured hegemony of capitalist growth. Echoing the demands of movements for agrarian and environmental justice that are strongest in the Global South, degrowth is a call from the centers of neoliberalism in Europe, North America, and Japan to reorder society and landscapes by "escaping from the economy" by countering the dispossession and exploitation of modern capitalist structures, processes, and institutions.[22]

Living differently, with greater solidarity for social justice and the radical equality of all living beings, will no doubt require transformation of nearly all societies, be they in the North or the South. Yet the movements and economies needed to dismantle dominant modes of economy and the state are still largely unclear.

A starting point for all efforts of liberation from the structures of the state, capitalism, and the corporation is to inquire how social movements have created emancipatory power to stop capitalist growth right now, in the actually existing present. Direct-action protests by Indigenous peoples, reclaimers, the landless and unemployed, and deep green environmentalists often shut down the infrastructures of capitalist dispossession like timber operations, plantations, mines, and oil and gas fields. These are disruptive resistance movements as well as movements for the creation of noncapitalist autonomous spaces. These movements have created new worlds where once dispossessed peoples control the land in collectives and many forms of life beyond the human regenerate and grow.

As the rest of the story of the Casiavera reclaiming movement will show, lived practices of degrowth exist, are central to our shared futures, and are not limited to reformist politics practiced in the hallways of democratic institutions. Generations of struggle in city squats and out in the countryside in the fields and forests have unfolded as a collection of related struggles for territory, resources, and ecologies. Machine-breaking, blockades, land occupations, street marches, and strikes constitute these modern, non-violent, direct-action movements.

These movements represent the greatest possibilities for decolonizing the land. The centrality of these direct-action movements to remaking life otherwise cannot be overstated, for to be alive and free is to be in collective movement with others in organic action.[23] In contextual ways, Indigenous and reclaiming movement communities in motion have already made degrowth real. Or to use the language of these social movements, they have struggled to make new worlds otherwise.[24] These worlds are territories where new relations can unfold across protests, communities, and landscapes, widening the possibilities of what life can be.

STRUCTURE AND SOURCES

Understanding the significance of Casiavera's reclaiming requires looking back into the troubled social and political history of the Bukit Barisan forests. Accordingly, in the first part of this book, "Dispossession," I begin my account with the very moment that New Order dispossession arrived in Casiavera.

I consider how control of the land in Casiavera swung away from small-holders and toward monoculture plantations at the start of the New Order under the barrel of a gun. I examine how colonial and New Order rulers constituted their authority across Aren with weapons, the law, land surveys, maps, and ideas of development and progress. Only with an understanding of Casiavera's colonial, capitalist, and industrial histories is it possible to appreciate the importance of Casiavera's subsequent reclaiming movement.

In the second part, "Reclaiming," I turn to how reclaimers established a mobilization movement centered on direct action, collective land control, and diversified livelihoods on the land, from the mid-1990s through the present day. I attend to the ways Casiavera's onetime plantation laborers countered the state and corporate control of the landscape with protest and land occupation, constructed uses for their reclaimed land in collectives, and elaborated an agroecology as part of their smallholder economy.

I came to live in Casiavera nearly two decades after reclaimers first occupied the land. My ethnographic research therefore draws on what are representations of movement history. These histories of protest, mobilization, and livelihood are all political statements, what David Graeber called "arguments about the ultimate meaning of protest."[25] Everyone involved in the struggle for land in Casiavera was attuned to the vital idea that how the movement was remembered would make all the difference for the land occupation's viability and the chance that subsequent reclaiming activists could move forward with their shared movement aims.

Recognizing that no one experiences social movements in their totality, I focused on perspectives that convey the transformative aspects of reclaiming. Absent perspectives include internal critique among activists because years into their movement, people told me they had moved on from that stage of their lives. What matters now is the way their movement supported their ongoing and ever-changing land reclaiming in the existing present.

As much as feasible, I pay close attention to individuals' varied roles, genders, and experiences. My account follows migrant laborers, skilled smallholder farmers, forest collectors, active participants in agricultural collectives, go-it-alone types, the manager of a mosque, and a recovering drug addict as they lived through dispossession and then reclaimed the land. Their experiences reveal the difficulty of regaining control of the land as both resource and home that allows for their survival. They also illuminate a path forward beyond the plantations and mines. Reclaimers' return to the land, and their uses of it, reveal smallholders' guts, energy, and creativity.

I write across Casiavera's broad history of agrarian change, even while I try to bring these ideas down into lived experience, centered in individual narratives. My study joins a line of research that has drawn insight into agrarian changes with the study of a singular plot of land.[26] A microhistory of Casiavera brings continuity to my wide-ranging discussions of ecological, political-economic, and cultural changes. While not a multisited ethnography, my attention is also not exclusively microscale: I take a wider gaze and focus on how the countryside, in general, is changing. I do so to not lose sight of the multiscalar forces that influence ecological transformation, class formation, and ideology. I follow a historical narrative, but it is not strictly linear, jumping eras and changing scales.

The historical turn in anthropology motivates my inquiry, reminded of Evans-Pritchard's humanistic, politically charged anthropology.[27] Historically rich inquiries are important because they allow scholars to speak to the very political processes of the production of culture and nature. Historical perspectives allow for the broadest kinds of social questions, such as, "How did society (and nature) come to be the way it is?"

I draw on an eclectic set of research materials, including historical documents, maps, and satellite photographs, oral histories, and walks across the landscape. I use my own photography as well, with the belief that long-term ethnographic work integrating text and imagery can bring together emotions, politics, and theory in powerful ways. My decision to include people's faces in a few of my photographs in this book was made after much discussion with these people, whose life stories are also included here. These reclaimers wanted to be shown without obfuscation, as testament to all they have built and achieved.[28]

To build a history of Casiavera's reclaiming, I worked for two months in the archives of Casiavera's local government offices, as well as the Indonesian National Archive in Padang and Payakumbuh, West Sumatra. Sources outside of Indonesia include the archives of Leiden University and the Hoover Institution, Stanford University. Most important to my study is my use of an archival source created outside the purview of colonial and authoritarian government archives: a unique archive of letters held by the Casiavera Community Council covering communications between the council and the state about the late 1990s land occupation.

During exploratory research of historical maps and satellite imagery of the Aren volcano, I was surprised to learn that the US Cold War–era Corona surveillance satellite had captured imagery of the volcano from space. With these satellite photographs, I connect Cold War geopolitics, technologies of surveillance, and the arrival of a predatory process of plantation expansion in Casiavera. I take a critical humanist approach to analyze these technologies' effects on society as tools of power and control while at the same time making use of these satellite photographs for their insights into how the landscape changed along with the political-economic fluxes of the last fifty years.

Mixed methods aside, I spent the bulk of my research time engaged in ethnography. Except where noted, I witnessed all the events featured in this work firsthand over many hundreds of hours of participant observation, when I lent a hand to work on the volcano. I accompanied women and men into the forest to cultivate their plots and went with them to collect, harvest, and hunt. I joined in transporting produce to market, accompanied government officials as they went about their land and forest surveys and agricultural extension meetings, and had conversations long into the night with peasant union members in their squats on contested lands. All these experiences allowed a measure of understanding of the reclaimers' experiences in a way that interviews or surveys never will. The slow rhythms, the physical hardship, and the never taking a day off from the farm took me deeper into life on Aren.

During my time in Sumatra, I moved fluidly between participant observation and ethnographic analysis. Rather than take ethnography to be a research method per se, during my years in Sumatra I came to think of ethnography as a way of being in the world. Ethnography is the art of careful listening and detailed observation. In the parlance of today's young people who tend to think a lot about our sociological problems, ethnography requires "checking your privilege." A way of ethnographic being demands that the ethnographer nurture an unceasing awareness of one's social positioning among hierarchies of power, capital, and marginalization. No matter how comfortable I became and how fully others embraced my presence, I tried to always consider how my experience was at once shaped by social phenomena and impacting others' lives in unanticipated ways.

Early on in my fieldwork, while I was traveling through Sumatra's Bukit Barisan mountains by bus, a young professional, his Blackberry smartphone in hand, questioned one of my closest collaborators, Nikman, about just what he thought he was doing talking to a man like me. "Leave him alone," he said as Nikman asked me what I would like to eat for lunch. To the man from the city, it was unimaginable that someone who looked like Nikman, with the weathered face, wiry build, polyester slacks, and rubber flip-flops of an agricultural laborer, would have any business being my guide, travel companion, and friend. Nor could the young professional consider that I, as a white, not-nearly as lean, American man, would have any business traveling with a rural worker like Nikman.

The tendency for many in the Bukit Barisan to misplace me was a recurring theme of my fieldwork. Many asked if I worked for agribusiness, worried my arrival signaled the expansion of the plantation estates into their homes. As my time passed in the central Bukit Barisan uplands, I developed friendships with smallholders who understood my work to be learning about their family's lives and livelihoods with the aim, as they put it, to teach Americans about practices of agriculture that were different from the kind done by industrial agribusiness.

Eventually I was invited to spend time in a range of forests, with a range of people, involved in a range of individual and collective struggles for survival and well-being. I mainly noted observations of mundane life during casual conversations as daily events unfolded. In the evenings and the heat of midday when people were not working, I noted oral histories. As a man in my thirties, older men told me on more than one occasion not to be seen talking with women alone. I generally did not do so, aside from a handful of exceptions when I felt my presence was welcome and the conversation easygoing. Usually my conversations with both women and men took place in groups, with many voices interjecting, in places like markets, fields, small eateries, and homes.

I sought to be open to what all kinds of people were up to, for ethnography is only useful when one is open to listening, observing, and feeling. Unfortunately these skills are not often learned or practiced by those in power, perhaps because the many compounding regimes of dispossession that global hierarchies are built upon depend on keeping a distance from those that rule and others who do not share their own lifestyle or perspective. But these are skills that we need more of in our society if we are to make it more equitable and just.

PART I

Dispossession

Roaring bulldozers uprooting trees mixing with screams that fill the jungle. The strange greedy clowns just laugh.

—From the 1980s folk song "Isi Rimba Tak Ada Tempat Berpijak Lagi" (Nowhere to rest for forest things) by Iwan Fals

CHAPTER 1

Under the Gun

After a few months living on the volcano, I received an unexpected invitation from Wibawa, a religious scholar and bureaucrat who had learned I was interested in the history of Casiavera's collective land. That same afternoon I made the trip down the volcano to meet Wibawa at his office in the city, where he managed a district public health program. In a dirt parking lot, he introduced me to his coworkers as a scientist studying agriculture. Close enough, I thought, coming from someone I had not yet met. Inside, while I was sitting on a worn couch in Wibawa's spacious office, he threw me off when, before engaging in any small talk, he asked to see my research permission and visa paperwork.

"You have it with you, yes? You are always required to have it with you. You have permission for your work, right or not? You are an official researcher, yes?"

I was surprised by the pace and precision of the questions. Although it had been weeks since anyone last asked for my paperwork, I did have it, in a blue plastic folder in my backpack. I handed it over. Wibawa's stream of questions continued even as he used his finger to scan through the letters' many lines of text.

"You have notified the district immigration post? The police? The village head? They all hold copies of these letters?"

"Yes, yes, yes," I nodded. "I am 'official' (*resmi*)."

Satisfied, Wibawa returned my blue folder to me. Then he reached into his own briefcase and handed me a three-inch-thick stack of black-and-white photocopies.

"Here it is, the history of our land, the way we lost it. And got it back. It is all here."

Flipping through the first few papers, I realized I was holding an archive of documents written by individuals, the Casiavera Community Council, and state representatives over two decades, mostly letters between various parties involved in the disputes that had arisen during the long effort at reclaiming Casiavera's collective land.

By then I was certain such an archive did not exist. My inquiries and searches in the state archives had turned up only cursory and generalized references to Casiavera. I had spent days in Casiavera's village head's office and meeting room, reviewing the stacks of documents and booklets there. But among the community development plans, budgets, meeting lists, and marriage, birth, and death registries, there was only the ubiquitous absence of any documents to do with the radical transformation of the land from community Indigenous collective land into a corporate cattle ranch and plantation, and Casiavera into a company town replete with a pool of wage labor. Even stranger was the lack of any mention of smallholders and the Community Council's long effort of reclaiming.

"But Wibawa, where did you get these documents?" I asked.

"What, you thought they would be in the offices?" Wibawa smiled. "Well then, everyone would know about them, wouldn't they?" he added.

Wibawa received the rest of my visit that early afternoon with grace. He smiled and nodded while I asked my questions. But he answered none of them. That day Wibawa told me only that he was from Casiavera, and, as evident from the stack of papers he handed over to me, for decades he had felt compelled to make copies of any documents he could get his hands on that were important to understanding the changes that were occurring on the volcano. Because these were his copies, he felt that he could share them now, with me, because I was a researcher in Casiavera with the official invitation of the state.

As a careful bureaucrat, Wibawa's deeper understanding of my positioning as a scholar of social movements went unsaid, but I knew that before Wibawa gave me his archive he spoke more than once about my work with Malin, the longtime reclaiming movement organizer who first hosted me in

Casiavera. My arrival with Malin to live in Casiavera was not happenstance. Kayat, a noted organizer in the Sumatran environmental justice movement, introduced me to Malin because Malin lived a life rich in reclaiming activism and direct action in Casiavera. And Kayat only made this introduction for me because starting in 2007 Kayat and I had worked together on campaigns to stop the expansion of industrial oil palm plantations.

Kayat saw how over the years my work shifted from US non-profit organization campaigner to a scholar locating my ethnographic research in streams of analysis committed to liberation struggles. With the goal of upholding activists' sacrifices while also spurring reflection and new social movement action, this research demanded that I take part in movement analysis and protest action, where I contributed to shared research framings and goals informed through cycles of mutual reflection.[1] I worked through activist research alliances, made my commitments to social justice and nonhierarchy clear, and kept up ongoing conversations with the people I worked with about how power relations structured my identity and work as a white cisgender man from California.

Perhaps most importantly, both Kayat and Malin understood that I worked to support global environmental agrarian and environmental struggles, moving between periods of protest participation and research in the United States and Indonesia. Although engaging in critical discussion of protest mobilization while also participating in it is always a high-wire act, my political engagements were the reason Kayat, Malin, Wibawa, and Casiavera's other reclaimers allowed me to work with them to share their story.

Wibawa and I came to spend many days together on the volcano over the coming months, but Wibawa never did speak with me about the meaning of these documents for him, or how he remembered his life during the time his archive details. His silence about his lived experiences of the New Order's dispossession of his community's land was nearly complete, even as he held the deepest of knowledge about its terrible effects. More important still, Wibawa never shared with me his insights into how to counter colonial and capitalist dispossession.

Still, Wibawa gave me the gift of his archive. This gift gave me a new fieldwork ritual. Most nights for months in Casiavera, after the evening meal I would open the folder and spread my own copy of Wibawa's archive across the floor of the house on the volcano where I stayed. The archive came to me already in chronological order, and to read through it during those nights on the volcano was an experience of crossing scales of time and space even while I remained centered in place on the floor of my house, immersed in the familiar sounds and smells of the deepening nights. If visitors joined me for these archival readings, as they often did, the discussions that spun out of the documents would often bring us up to Isha'a, the day's final call to prayer, without having made our way through more than a few pages.

Circling and redoubling through the dates, details, and language of the archive allows me to distill Casiavera's story of the very day that their New Order dispossession began at the point of a gun.

ONE MOMENT OF DISPOSSESSION

It was June 1968 when the village head of Casiavera, recently appointed by the New Order regime, called a meeting with the Casiavera Community Council of elected Minangkabau family-lineage representatives. A few days later, traveling down the volcano on horse and by foot along a dirt track that connected their homes, the council gathered in a mosque at the base of the Aren volcano, which sits at what is still the boundary of their land. A handful of armed soldiers, the district head, a retinue of bureaucrats, and a businessman named Mahmud Teuling met the smallholders there. Once everyone was settled cross-legged on the tile floor, the district head opened the meeting. He told the assembled community leaders that the national Agrarian Directorate in Jakarta had signed over their collective land to Mr. Teuling, a retired military police general. Casiavera's smallholders were to immediately relinquish their collective land to Teuling.[2]

Murmurs of dissent spread in the mosque that morning. The village head (kepala desa), trying to ease the sense that he had betrayed his community, said that he had been given the news suddenly, without a chance to object. Casiavera's representatives pointed out that the land was their property. Not only was it theirs, but the land provided livelihood to people living on the upper reaches of the volcano. The fact that they were called to the meeting at all proved their point, they said. Why would the district head need to have this meeting to tell them to give up the land if the land did not in fact belong to them?

Slamming his hand down on the floor in front of him, the district head silenced the growing chorus of dissent, telling the room that anyone who disagreed must be Gestapu, the Indonesian Army's term for a movement of supposedly violent communists seeking to overthrow the young republic.

Gestapu became an army watchword for evil after a clique of soldiers kid-
napped and killed six army generals in 1965, dumping their bodies into a
well in Jakarta. Without evidence, anti-communist army leadership accused
Indonesia's Community Party of directing the assassinations of the army
generals as part of what the army called the "Gestapu movement."

Defending the archipelago from the manufactured threat of the Gestapu
became the army's pretext for their mass murders and detentions, designed
to dismantle Indonesia's thriving leftist and non-aligned majorities, in the
process bringing forth Indonesia's authoritarian New Order, a thirty-two-
year military dictatorship. With Gestapu, the army created a term that
equated leftists and dissidents with the depraved Nazi secrete state police,
the Gestapo. The term was an attempt by the army to make the Indonesian
people feel as if their morality and safety were under threat.[3]

The idea that Casiavera's family lineages would be labeled Gestapu
brought a wave of fear to the group at the mosque. The threat came just as
the New Order's terrible pogrom was coming to an end. The regime's com-
mandos and their paramilitary squads still roved the cities and countryside,
murdering anyone they fingered as Gestapu. At least five hundred thousand
communists, labor and peasant activists, feminists, scholars, students, artists, and minority ethnic groups, among many, many others, were murdered.
More than a million other people were incarcerated.[4] Those living on Aren
had intimate experience with this violence. West Sumatra was one center
of the killings.[5] Nearly everyone on Aren could be connected in one way or
another to agrarian leftism through their ongoing squat on the colonial-era
W. H. Samuel ranch that had begun two decades earlier, after the Japanese
invasion of World War II. Everyone in the mosque that day knew that the
district head's threat to label their community Gestapu communists could
be a death sentence. Casiavera's leadership sitting in the mosque also knew
they had to be especially careful, because leftist politics had long circulated on the volcano. Indeed, Indonesia's most influential communist, Tan
Malaka, was born less than a day's travel from Casiavera.[6]

The Casiavera leaders were chilled into silence. The meeting did not last
an hour. According to one of the heads of the family there that day, it was
a "lightning fast" deliberation. The community leaders made the hour-long
walk back up the volcano along a muddy track to tell the others: the government had seized the right to their land and turned it over to an agribusiness
company.

FIGURE 5. Two photographs of the owner of Dona Company, Mahmud Teuling, as a young military policeman.

Mahmud Teuling arrived in Casiavera a few weeks later. He first gained a regional reputation when he commanded a small attachment of military police during the early years of the Indonesian Republic. During the first decades of the New Order, he rose to two-star lieutenant general. In a surviving photograph of him, a young Teuling looks at the camera with an easy, piercing gaze. He looks surprisingly young in the photograph, dressed in his army uniform with a Military Police pin set neatly on his collar.

Reclaimers in Casiavera remember Teuling as the ultimate authority on the land. Up on the volcano people recalled the power and bravado of he whom they interchangeably called "Mr. Dona," "Mr. Army," and "Father Teuling." An older Chinese Indonesian couple born in Casiavera, Kuman and Vania, recalled for me how Teuling would ride his white horse along the track that marked the concession's upper boundary with the cloud forest above and also lower down the volcano along the cobblestone road he had built to mark off his land from the village of Casiavera.[7]

Both Kuman and Vania were middle-school students at the tail end of Teuling's reign. Kuman left Casiavera when he was seventeen. Down on the coast, he and his five friends snuck onto a ferry to Malaysia. He worked there for one year. Back then, he says, it was not hard to get across the Malaysian border without documents. He worked cutting and hauling oil palm fruits in two huge oil palm plantations outside of Negri Sembilan and Johor. Back on the volcano he met Vania, who had been living at home working mostly in her family's rice paddy, when he was out on one of his late afternoon motorbike rides. They joined other reclaimers to recall how Teuling was the "master," who once "dominated" (*menguasai*) life on the land.

In Wibawa's archive there is no record of any protests against the company's arrival. Unlike during the occupation of the plantation that would come thirty years later, there are no surviving letters from concerned individuals or the Community Council objecting to the opening of the company's livestock ranching operation during the New Order. There are no records of trips down the volcano for meetings with government officials in an attempt to sway them. Nor did I hear anything about street protests or other direct action in my conversations on the volcano. It seems that there was mostly acquiescence.

Under Major General Suharto's New Order, power was not invisible or abstract; it was made material through government and civilian actors. Power was the force that brought some to arrange death squads, compelled others to participate in genocide, and caused the murder and internment of millions. In the heady killing fields and gulag, and later in their all-too-durable aftermath, power was the ability to choose to participate. This power was reserved to the few elites. Those in Casiavera did not choose to accept or reject the land concession and Dona Company; dispossession was imposed upon them with intimidation and force.

Even still, a few of the most determined smallholders chose to test the state's resolve to enforce Mahmud Teuling's land control. After Teuling began to clear the land, one man continued to walk up behind his home and cultivate his sugar cane crop. Police found him in his field on the land and beat him badly. Predictably, the beating sent a chilling message to others: Mahmud Teuling was the master of the land, and he had all the backing of state power behind him.[8]

In Casiavera, Teuling's personal connections to the New Order regime were said to have gone all the way to Suharto. After retiring, but before acquiring the land on which he built Dona, Teuling became wealthy by capitalizing on his New Order connections. With a difficult-to-acquire tobacco business permit, he became a major trader of pre-rolled cigarettes. These same personal connections allowed Teuling to secure the "contract," the cultivation rights (Hak Guna Usaha, HGU) to the land above Casiavera.

A New Order legal instrument used for transferring control of land to corporations, the HGU was a state land lease, or land concession, similar to the Dutch colonial-era *erfpacht* land concession that had codified the first waves of plantation expansion across the archipelago in the early 1900s. A foundational technology of rule and control of space, the New Order HGU reconfigured Sumatran landscapes in the image of international agribusiness and resource extraction companies. The Dona Company's state land lease declared smallholders' private and collective land to be state land, under the control of the company.[9] The state's creation of the concession brought a form of capitalist land enclosure and control by dispossession to Casiavera. There was significant power and material effect in the HGU as an object in itself. What Kuman the reclaimer told me about it mirrored comments by scores more: "It was an HGU, for industry. Our Council could not do anything about it. It was an official government contract."

Mahmud Teuling's military position allowed him the standing to petition the government for a land concession. With it the military man restarted the Dutch colonial-era corporate control and exploitation of Casiavera's land that ended with the Japanese invasion of Sumatra and subsequent expulsion of the Dutch East Indies colonists during World War II. Teuling's profits in trading cigarettes gave him the financial capital to immediately begin transforming Casiavera's peasant forests into grazing land. The differential power constructed and sanctioned with the land concession—a form of extra-economic inequality—brought the Dona Company ranch into existence where peasant forests once grew.

The establishment of a land concession at Dona ties Casiavera into a wider process of agrarian change that was to sweep across New Order Indonesia. Casiavera's loss of their customary land unfolded within a field of social conditions of neocolonial capitalism that made modern industrial agriculture possible in the Global South. In a complex story, anti-communist Cold War military interventions, the advancement of technologies of surveillance and rule, and the consolidation of a high modernist authoritarian regime brought powers of persuasion and exclusion to bear on the volcano.

Agrarian violence bled into the New Order land enclosures. Just months before Casiavera's collective land was lost, the military conscripted young men on the volcano to kill communists. As one man told me, "I was at the front with a machete and the military was behind me with their guns. . . . The Communists were anti-God. They were hanged (*gantung*), eliminated (*bubarkan*), thrown to the gutter (*buang talang*). . . . They were brash (*kurang ajar*) and savage (*ganas*)." Who benefited from the violence and who was expelled or injured had little to do with people's actual participation in leftist mobilizations. Survival depended much more upon local political affiliations, social standing, and luck. A military dictatorship's arbitrary violence pervaded state-making, allowing industry to dispossess Indonesians from the land and radically transform the landscape.

LAND DOCUMENTS, MAPS, AND CONCESSIONS

Wibawa's archive contains a series of unique state agency letters that detail how the state enacted and exerted the power required to dispossess smallholders from their land. Most important of all was the land use permit (Surat Izin Pemakaian Tanah) from the provincial Agrarian Inspection Division. Multiple state authorities signed the land use permit as the first of several documents required to authorize the final land concession letter, or "right of exploitation" (Hak Guna Usaha).[10]

The October 1968 land use permit's simplicity belies its influence. It is a typewritten letter that defines the plantation's boundaries without a map and with only general reference to the land's north-south and east-west boundary locations. There is no statement of the plantation's total size. The land was decreed to be for the exclusive use of Dona Company, and it would be up to the company to establish its boundaries. Written in the name of the governor of West Sumatra, the permit gave a few terms of the land transfer.

The land in question was defined to be state land. It was to be leased, not sold, to Dona for twenty-five years. In response to a previous letter Dona staff wrote to the Agrarian Inspectorate, the governor states that the head of the Agrarian Inspectorate was swayed by Dona's request and Teuling's intentions to establish a "livestock and agriculture corporation."

To support the company's land concession permit, the agency also issued a decree of rights letter (Surat Keputusan Pemberian Hak), also called a land book (Buku Tanah). It is a straightforward register of the land concession number, year issued, district location, and signature of approval of the presiding bureaucrat in the Agrarian Directorate. Its most noticeable feature is the Indonesian state seal prominently affixed to the center of the cover.

Like the 1968 land use permit, the land book is a textual projection of state sovereignty and authority. These documents act in a way that allowed the state to strengthen its institutional power and convey how its bureaucrats believed the land should be used. With just a few black lines typed on a page, these two state documents provide only the faintest guide to how new forms of power would be inscribed onto the land, but they formed the basis from which this new, radical power emerged to control Casiavera's land and people. For the recipient of these documents, company owner Mahmud Teuling, they provided the justification for taking total control of the land.[11]

In many ways the New Order's dispossession of smallholders, creation of land concessions, and handing over of the land to industrial corporations was a return to colonial modes and logics of controlling land, resources, and people. At its core, the land concession is a primitive form of colonization: violent dispossession by enclosure. Preceding the New Order land concessions, Dutch colonial administrators constructed a legal framework to bring into being the large-scale land concessions that dispossession requires.

Before the Dutch established a colonial concessionary capitalism based on the state allocation of land to multinational agribusiness and resource extraction corporations, they fought a series of wars with the Indigenous polities of the archipelago to establish colonial rule. These wars of occupation began in the 1820s in West Sumatra with an invasion of the Minangkabau territories, known to be among the wealthiest Indigenous polities of the time. From there, the colonist armies expanded their wars of occupation to Java, Borneo, and Sulawesi. The Dutch would wage war up through the turn of the century, eventually claiming tenuous control of nearly the whole of the archipelago by 1910.

After their occupation of Sumatra, the Dutch signaled their intention to establish a new form of colonialism based on wide-scale land dispossession with the 1870 Agrarian Act (Agrarische Wet). The act established the state's legal authority to turn over Indigenous lands to European and North American agribusiness through a legal tool known as a land concession, or *erfpacht*. In 1874, the Sumatran Domain Declaration followed up on the Agrarian Act to specify how the enclosure movement was to unfold in the Bukit Barisan using the principle of domain declaring (*Domeinverklaring*), a form of eminent domain (i.e., enclosure and dispossession) the

Dutch state invented to codify new *erfpacht* land concessions for plantation monocultures.[12]

The *erfpacht* as an instrument of dispossession was constructed out of earlier Dutch experiments with land control.[13] After generations of unrest, uncertain rule of the archipelago, and declining Dutch East Indies revenues, the colonial administration concluded that a new, central authority was required to control how land itself was used. Indeed, control of the forests; their annihilation; and their replacement with plantations, logging operations, and mines became one pillar of colonial and post-colonial state rule.[14] With their spatial and temporal heterogeneity, peasant forests were difficult to monitor, tax, and, ultimately, exploit.

The Dutch had come to think of smallholders' farms, forests, and the commodity crops they grew there as the "robber economy" precisely because they in part remained outside of colonial control. And because they existed outside of colonial control, the Dutch colonists considered them a form of theft from the state.[15] The *erfpacht* land concession was the way the Dutch would transfer control of the peasant forests to the growing transnational agribusiness. These corporations promised to be a more loyal colonial collaborator than the Indigenous polities. And with their large-scale, capitalized estates, northern agribusiness would be easier to tax, monitor, and discipline.

The land concession is unique for the way that it can inculcate state discipline, demarcation, and exploitation from afar. It was an invention of the highly centralized European colonial states to control outlying geographical areas. Concessions extended the power of these states across varied social contexts and environments by bringing in the capital and power of corporations. Concessions did so as conceptual and legal constructions that fused the taking of land with the making of a distinct form of state-owned and corporation-controlled property.[16] As such, these colonial concessions were fundamental tools of elites that sought to profit from the processes of dispossession that transformed entire regions. In both colonial and post-colonial contexts, land concessions underpin epistemological boundaries as well, serving as zones of social order, simplification, and standardization, all as justification for the belief in the superiority of Western scientific forms of knowledge.[17]

The *erfpacht* was a social projection of control that required spatial knowledge. Increasingly, the Dutch invested in mapping technologies and land surveys to create their *erfpacht* and determine legitimate control of them.

But up until 1850 the colonists had yet to develop the kinds of spatial science like cadastral surveys and topographic maps that land concessions require.

Take the 1842 map of Sumatra's central Bukit Barisan Mountains that P. J. B. De Perez hand drew from his surveys of the island from 1820 to 1837.[18] De Perez was among the first colonial employees to map Sumatra's uplands, which were not yet under complete Dutch control. The map captures upland geography, rendering the four volcanic peaks that rise above West Sumatra's upland plain. It accurately places Sumatra within a cartesian grid using a scale of 1: 2.27 million and employs accurate relative cardinal directions.

Yet the landscape remains mostly unmarked space, still not surveyed and so unknowable to European agribusiness and Dutch administrators writing the *erfpacht* land lease letters from their offices in Batavia. In fact, the scale of the De Perez map made it of little use for anything except long distance navigation. The colonists going out into the uplands in search of specific landscapes to remake as *erfpacht* long leases required a more detailed, fine-grained spatial knowledge.

A step toward that end was the first topographic map of the West Sumatra uplands, drawn in 1855 by A. J. Bogaerts, an active cartographer for the Dutch government in Batavia, from field data collected in 1854.[19] The finer scale allows accurate relative location and elevation of individual mountains, watersheds, and plains. Bogaerts drew these in a precise black ink shading. Elevation information was integrated with road and community locations to, for the first time, give Dutch planters an image of uplands geography. Production and distribution of these maps brought the Sumatran uplands into clear focus for colonial planners, capitalists, and the military. As new spatial technology, they gave the Dutch both a rational and performative tool to govern the countryside.

After production of Bogaerts's map, forester J. W. H. Cordes recorded the first Dutch survey of West Sumatra's upland forests in 1874.[20] In the employ of the colonial administration's Forest Service, Cordes recorded the location, extent, and uses of the forests against Bogaert's 1855 map. While far from a panopticon of land control, these cartography and forestry survey sciences together represented newfound capabilities in spatial technologies of power. Sitting side by side on a desk in Batavia, the map and land use survey became a compelling diagram of upland geography and how people used the land.

Shortly after Cordes began surveying the uplands, the colonial government founded the Batavia Topographical Bureau. The service integrated existing maps and land use surveys to produce a new series of maps of the uplands that included accurate depictions of agriculture and ecological habitats at a scale of 1:20,000. Details like grasses, banana and sugar palms, forests, footpaths, roads, rivers, stone buildings, and graveyards were given, while contour lines gave form to landscapes. From 1886 to 1902, cartographers produced thousands of systematic, hand-drawn maps of strategic sites within the colonial territories. All these maps could be indexed against a large-scale national map to give their relative locations. Using this index informed understanding of entire regions from colonial offices in Jakarta.

The uplands became mapped and thus known to Dutch administrators from afar. Attempts at colonial dispossession and control quickened, supported by the new ability to visualize Sumatra's agricultural potential. Indeed, cadastral surveying and mapping became instrumental to colonial suppression of smallholder ways of relating to the land that rejected ideas of land as a resource to be exploited. In a process of dispossession, with their land concessions Dutch colonizers gained control of Indigenous lands while also recoding the meaning of land into an abstract legal entity. For the first time, widespread dispossession by enclosure unfolded across the Bukit Barisan.

In the hands of the Dutch, maps and the *erfpacht* they made possible quickly became legal justification for a state-led enclosure movement against Indigenous peoples' rice terraces, agricultural fields, and fallows, and even larger expanses of upland peasant forests. The cadastral survey and map ended attachments to land based on lived social experience, replacing these relations with ownership of land signified through documents created in government offices.

The colonial map and land concession survived the end of colonialism virtually unchanged, outlasting European domination and becoming central to the massive expansion of the land concession plantation, logging, and mining system of Indonesia's New Order military regime in the second half of the twentieth century.

Not necessarily oppressive or emancipatory, a long series of state rulers deployed maps and land surveys as spatial technologies for collecting and communicating large amounts of social and geographic knowledge across distance. Yet in practice these spatial technologies were created and used to

visualize and enforce forest reserves and plantations, mines, and other extractive territories. As such, these maps were instrument of spatial authority
and territorialization, meant to erase historical Indigenous uses.[21]

Across centuries, the authority found in maps and land concessions was
real but incomplete. As the New Order moment of dispossession of Casiavera's peoples revealed, it was the state-directed violence found at the end of
a gun that forced smallholders' acquiescence to these spatial tools of control.

CHAPTER 2

Primitive Enclosures

Before the Dutch dispossessed the Aren Minangkabau of their land, Dutch colonists sought to dispossess them of their labor. Taking up the French concept of forced labor, or corvée, Dutch administrators and soldiers used violence and coercion to remake nineteenth-century smallholder and forest life with their colonial efforts at forced coffee cultivation. It was colonial exploitation through and through, aimed most directly at people's work and the spatial and temporal patterns of their lives.

With the onslaught of oil palm and rubber plantations that would directly dispossess people of their land still to come, the forced cultivation of coffee required nearly all Sumatran communities to produce coffee—without pay. Forced cultivation fundamentally altered society, bringing increases in the circulation of trade, capital, and currency in Sumatra's uplands. For a short while it made some colonizers rich, as well as a number of Sumatran landholders and traders who were able to profit from the coffee trade despite the best efforts of Europeans to shut them out of it. At its end, forced coffee cultivation brought inequality and ecological collapse.

Sumatran forest farmers' history with coffee is so deep that many Indonesians consider the crop to be native to Sumatra, although it is not. The Arabica coffee plant did not arrive in Sumatra until the end of the seventeenth century, by way of pilgrims to Mecca, who returned home through Yemen with smuggled seedlings.[1] Around the same time, the Dutch East Indies Company (Vereenigde Oost-Indische Compagnie) handed out Arabica seeds, first in Java and then in Sumatra, to spur the crop's cultivation after they took control of the Arab coffee gardens of Sri Lanka, in 1658.

Seeking direct control of the coffee economy, in 1839 the Dutch governor of the West Coast of Sumatra, Andreas Michiels, among the most ruthless of all Dutch administrator-colonists, wrote the Ministry of the Colonies to inform them of how he would implement his plan to transform Sumatran society with forced cultivation. Instrumental to the system was colonists' use of Dutch-sponsored village chiefs to act as "superintendents of agriculture."[2] The chiefs were created as an Indigenous extension of Dutch rule to be native lords, labeled aristocrats, and responsible for enforcing corvée. Dutch lawyers and administrators drafted a set of policies to bring the system into law that aimed to gain a stranglehold on the coffee economy across the Sumatran uplands. New Dutch coffee trees were to be grown as a kind of smallholder plantation, on which households were forced to clear cloud forests, plant coffee, and bring in harvests without compensation.

Resident of the West Sumatran highlands C. P. C. Steinmetz issued orders to make all new coffee planted in monoculture lines and rows. New plantings were to follow what the Dutch called the Javanese model, following the small-scale monocultures developed as part of early forced cultivation on that island. The administrator considered the already existing Minangkabau coffee forests—diverse agroforests—poor agricultural technique, too wild and open to predation by pests.[3] The coffee plantations were a result of European settler-colonial sensibilities of agriculture, Versailles-like, wherein power and profit were inscribed into the straight lines of coffee trees, all the same age, that would eventually come to stretch across the upland's volcanos. The monoculture schema expanded through the Dutch's conquest and increasing control of agrarian places. It was a totalitarian vision of the use of space. The colonial profits accumulated were immense; the changes to the landscape were profound.

On Aren, leaders were forced to organize their own families to clear the lower bounds of the cloud forest above their communities. Where previously smallholders freely incorporated coffee into diverse agroforests, Europeans and their Sumatran collaborators carried out the regular inspection and survey of the land. They forced smallholders to grow in monoculture, terrace the land, and clear it, leaving the young coffee trees open to the sun, even though coffee evolved as a shrubby tree of the forest-canopy understory and is healthiest growing was in the shade. Dutch *controleurs* were

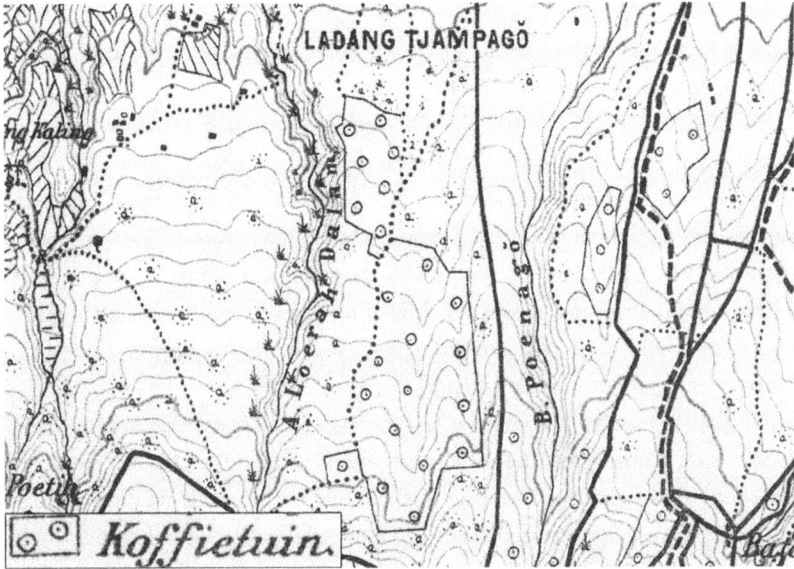

MAP 2. Casiavera's colonizers forced coffee monocultures across the landscape, as seen in an inset from an 1895 Batavia Topographical Bureau map. The central portion of this inset shows the land above Casiavera to be completely under coffee cultivation by 1894, labeled "Ghosts' Field" (Ladang Tjampago). The box at the bottom left of the image is the map's legend entry for coffee cultivation (*Koffietuin*).

the government architects and overlords of the system. Day-to-day production was managed by Indonesian coffee inspectors and collectors (*mantri* and *kolektor*). They checked the coffee plantings and village deliveries of the crop. Where they identified potential expansion areas, they demanded new coffee gardens be built.

An 1895 map from the colonial Batavia Topographical Bureau displays forced coffee's dramatic remaking of Casiavera's lands. The sheet depicts coffee covering everywhere from around homes at "Bunker Rock" up to the "Ghosts' Field" above the village bordering the cloud forest on the steepest upper reaches of the volcano.

Protests against the forced coffee system's exploitation and landscape changes were tenacious and ongoing despite Dutch coercion and compulsion. Workers sought to evade control at every turn. According to Dutch officials, West Sumatran uplanders were "obstinate" and "circumvented any regulation whatsoever." Police rolls were filled with charges of failure to perform "required labor duties," a situation "without equal in the Indies."[4]

One act of subversion was to simply not plant coffee and hope to not get caught. Migration away from the uplands became more appealing. Smuggling became a favored profession.

In West Sumatra, Minangkabau traders continued their pre-colonial trade in coffee, now as part of the underground economy. They walked up through the deepest forests of the Bukit Barisan to load their coffee on canoes on the eastern flank that descended the great rivers of Sumatra's wide eastern alluvial plain and flowed on to the ports along the Malacca Strait, where Dutch control was weaker. Living a life underground as a smuggler had the added benefit of avoiding the corvée demands imposed on those listed on village population registers.

By the 1870s, smallholders had largely abandoned the coffee monoculture. Left mostly unmanaged, the coffee plantations were vulnerable to disease. In the 1880s they were destroyed by leaf rust. This ecological weakness in the monoculture model provided an opening for Indonesian smallholders. Probably through Arab traders, around 1900 Sumatran smallholders acquired Robusta coffee seed, impervious to the fungal infection that affected the colonial Arabica plants. The new cultivar sparked the rebirth of coffee production, this time planted by autonomous smallholders.

The collapse of colonial forced cultivation and growing peasant coffee cultivation autonomous from colonial rule spurred the Dutch to create a new system of political-economic control in Sumatra. For the first time, colonial administrators sought to take control of the land itself, so that they could hand over massive concession areas to companies for exploitation as logging estates, mines, and plantations.

Dispossession by enclosure was made material in Casiavera on December 8, 1905, when the resident of the West Coast Division of the Dutch Indies government created the first land lease (*erfpacht*) on the Aren volcano. It was directly on top of Casiavera's collective land. According to the land concession letter, it covered the "cadastral division, known by the name Rimbo Sikadoedoek."[5] A German agribusiness, W. H. Samuel, had arranged to acquire the concession from the Dutch Council of Justice in Padang. The company targeted the Rimbo Sikadoedoek forest's valuable timber, and after clearing the forest, its rich soils. The letter gave W. H. Samuel "the right to stronghold" to log the forested flanks of the volcano and establish a cattle station on the cleared land. With such rights in hand, the company blocked access to smallholders' lands, including farms, forests, and pastures.

A decade later a Belgian-owned agribusiness, Halaban Plantation Company (Cultuur Maatschappij Halaban), took over the W. H. Samuel Company as part of Halaban's push to find land to plant chinchona and tea plantations on the volcano.[6] In November 1918 a Dutch assistant resident notified the governor of West Sumatra that during a meeting with seven community councils on the Aren volcano, Minangkabau family-lineage representatives had approved the Dutch government's plan to transfer this concession to Halaban and expand it by an additional 1500 *bouw* (7096 m^2) of land. According to the assistant resident, there was "more than enough land about," and community representatives had given their "full approval."[7]

One year later, in September 1919, Dutch judge commissioner Johan Dirk Pyper of the Council of Justice in Padang signed the land lease over to Halaban Plantation Company in the name of one Karl Baumer.[8] By the time the Dutch were in the position to accomplish this dispossession by enclosure and eviction they had spent nearly two centuries imposing themselves on Minangkabau society. It was not until the 1820s, however, when weakened by a civil war between supporters of the established matrilineal polities, called *nagari*, and a group of elite *shari'ah* (*sjarak*) Muslim insurrectionists, the Padri, that the Minangkabau submitted to Dutch colonial rule.[9] Immediately thereafter, the Dutch dismantled the Padri and went about restructuring the surviving but fragile *nagari* matriarchates.

During the formation of the Minangkabau kingdom in the sixteenth century, family lineages organized themselves into a confederacy of communities (the *nagari*) that functioned as autonomous village republics. Women's positionings and practices were compositional of these polities. Women elders controlled the home and land. They determined farming and other economic practices. And they organized life-cycle ceremonies.[10]

Geographic boundaries defined each polity's borders as a form of bioregional land management. On the Aren volcano these were individual watersheds, marked by the steep ridges that separated each of the many river valleys that formed in the cloud forests. At each of these bioregional polities, a deliberative, consensus-based council made up of elected family-lineage representatives, all men, arbitrated disputes.

Ultimate authority in each family lineage was separate and above the titled men of the council, wielded by a single elite senior woman, the Bundo Kanduang, chosen through deliberations of family-lineage women. Before the Dutch began their conquests of the Minangkabau in the 1820s, these councils and their elite women leaders were the highest forms of authority recognized in the polity.[11] Fundamental to women's control was the council's role in upholding prohibitions on the sale of land and sanctioning only the loan and inheritance of land within matrilineal family lines.[12] By the time of the crux of colonial conflict in the nineteenth century, the *nagari* were polities founded upon interdependent cosmologies drawn from Sunni Islam and Minangkabau matrilineal law and philosophy.

For many Dutch administrators, the villages were far too egalitarian and democratic for their purposes of conquest and exploitation. In the 1820s the colonial government divided the *nagari* into new administrative units, attempting to push the women family-lineage leaders into a merely symbolic role by replacing their authority with a new, all-male collection of village and regional headmen. [13] Seventy years later, in 1915, the colonial state redoubled its efforts to destroy the matrifocal village republics by appointing new male leaders with the powers of a feudal lord, not least the authority to collect taxes from the family lineages.[14]

The invented male authorities weakened but did not destroy the Minangk-abau matriarchate. The senior women representing each of the matrifocal family lineages, their longhouses, and their lands survived and resisted the Dutch's patriarchal rule as a center of local authority. The *nagari* and their family lineages survived the conjoined rise in the global commodity trade in the eighteenth century and colonialism with at least a measure of autonomy, but as the male Minangkabau signatories of the European agribusiness land concession in Casiavera in 1918 so clearly demonstrate, the decades of colonial pressure altered the *nagari* and their matriarchate's control of the land and society in elemental ways.[15]

Two decades after the Dutch Indies government created W. H. Samuel's 1905 land lease, forestry officials seized the cloud forests above Rimbo Sika-doedoek in 1923 as a state forest reserve. A portion of this land was not forest at all; it had already been cleared for timber and coffee.[16] In 1927 Dutch forestry official J. H. Chaping noted that many areas of these state-controlled lands on the Aren volcano included logged forests and grasslands. Feeling the urgency to control and exploit the shrinking forests, Chaping called for more demarcation and regulation of the boundaries of the forest reserves. On Aren, a Dutch forestry team built a wagon track to define the upper boundary of the W. H. Samuel cattle ranch and the lower boundary of the state forest reserve, an area of ancient-growth cloud forest. The Dutch colonists called it the Boschwezen Road, or Forest Spirits Road.

The Dutch moved to mark the boundaries of state forests in regions where population pressures on the forests were highest. In forest reserves and *erfpacht* across the West Sumatra uplands, Chaping directed local contractors to clear narrow inspection paths. Along these paths they planted *lenjuang*, a distinctive red-leaved shrub, to mark the *erfpacht* and forest reserve boundaries. Red-painted iron discs were also nailed to a boundary tree every forty meters.[17]

While the state ratcheted up its exclusion of people from the land, the countryside was becoming ever more densely inhabited. Up until the early 1900s the extensive topography of Sumatra, especially in the uplands, provided open land available for homesteading for many generations of Bukit Barisan smallholders. But these land settlement frontiers were rapidly closing under the expansion of the *erfpacht*.

In Sumatra, agro-industry erupted with the planting of *Para* and *Ficus elastica* rubber trees along the east coast plantation lands. In 1903 there were only some one hundred thousand rubber trees planted in Sumatra. By 1910 there were well more than three million. Oil palm plantations followed closely behind, starting along the west coast of Aceh in 1911 when Adrian Hallet opened the heavily guarded gates of the Belgian company Socfin's plantation in Indonesia.[18] A reiterative process among the plantation managers, finance capital, violence, and other technologies of rule spurred the emergence of the relations of production and power that came to be plantation capitalism. Foremost a territorial power, this plantation capitalism was founded on the Dutch's ability to force smallholders and other Indigenous peoples off the land and interject themselves into the organization of peoples' work and lives.

Enclosures to create the new colonial "forest lands" joined the plantation dispossessions. By 1923, 35 percent of the total land area in West Sumatra was state-claimed "forest land" that was often in practice home to lively agrarian villages. In 1927 the Dutch Forest Service estimated central Sumatra's extant forests to be over five million hectares, some 60 percent of the land area. Three million hectares of this land was already designated as state land, with plans in Batavia to approve an additional one million hectares of concessions.[19]

Depeasantization followed these large-scale land dispossessions for agribusiness, forest reserves, and mining across the upland. Local economies were greatly affected, with cultivation, grazing, and forest access—all activities that were mainstays of the upland economy—curtailed. Even the Dutch architects of the new concession system acknowledged it was a time of "shock," with inhabitants of the countryside facing hardship and damage.[20] The wave of enclosures pushed smallholders into a crisis of reproduction as land became more difficult to access, credit scarce and its terms often onerous, and wage labor opportunities few and poorly renumerated.

Many smallholders did not acquiesce to the colonial enclosure movement. As soon as the Dutch went about issuing their *erfpacht* land concession permits to European timber and agricultural corporations, administrators encountered resistance. Land dispossessions in the West Sumatran highlands in and around the Aren volcano pushed disaffected smallholders to form peasant unions. Nearly everywhere in the region Dutch agribusiness efforts at control were "hampered by the opposition from the local population."[21] By 1915, as the Forest Service went about its demarcation of *erfpacht* and the transfer of control of the land to European corporations, Dutch foresters wrote about the increasing tensions on the Aren Volcano:

> Locally, many problems have arisen with the local people, concerning the location of the forest boundaries. Also, the issue of long-lease parcels, for which advice is requested from the Forest Service, is giving rise to many difficulties, as there are hardly any maps, and the survey of the forests has just started.[22]

In many instances, smallholders refused to recognize the existence of these new state territories at all, a position made easier by the fact that the number of Dutch agents remained few. At the end of 1915 there was only one forester, three chief overseers, and two apprentice chief overseers in the whole of the West Sumatran uplands. These were joined by twelve Indonesian overseers (*mantri*), nine forest guards, three clerks, and one draftsman.[23] Chaping and the rest of the Forest Service experienced a fundamental precept of inhabited forests, in which claim and counterclaim are the conditions of forest life. Property rights beget power, and for that smallholders and colonial authorities were willing to struggle.[24]

As Forest Officer Chaping attempted to inscribe colonial boundaries across the forests, the locals felt the reduction of their livelihood possibilities. According to Chaping, "Locally this has led to serious problems."

Most difficult were the locations like Casiavera, where locals' land access was limited by the imposition of both an *erfpacht* and forest reserve in one landscape.

Already in 1911, the governor of West Sumatra, J. Ballot, announced there was insufficient good-quality land for peasants to plant because so much was granted to concessions. Governor Ballot summarized an alternative vision of development in his *Agrarian Plan and Regulations for Sumatra's West Coast*.[25] In it he argued for the complete recognition of customary territories, including all types of land under indigenous claim, predicting plantation expansion would cause a rebellion. Ballot was immediately concerned with preventing a repeat of the bloody 1836 tax revolt across West Sumatra, in which uplanders had assassinated several tax collectors and threatened full-on armed rebellion. In this history Ballot saw smallholders' revolutionary potential. And Ballot recognized that the *erfpacht* were destined to enact far greater changes than the income tax. But for his dissenting views the colonial government sacked and shipped Ballot back to Holland.

After Ballot was dismissed, Forest Service officer H. J. Kerbert wrote an objection to the effects of the *erfpacht*. He believed that smallholders had what he called "harvesting rights" (*plukrecht*) to the land that was being turned into land concessions. Kerbert recognized the centrality of smallholders' concepts of usufruct rights as opposed to individual ownership of land as property. In Kerbert's perspective, a system of land control that taxed smallholders but left their agricultural forests under their daily use and control would make for more peaceful forests. He proposed a state commission to develop a system of land control that could serve all interests, including Sumatrans'.[26] Kerbert was joined by an influential ally far above him in the hierarchical Dutch colonial service. In 1928 the resident of Tapanuli, just north of West Sumatra, dissented against the land concessions. The resident took the unpopular position that the peasant forests and the profits they produced belonged to the people who cultivated and harvested them. The concessions did not respect customary regulations or allow for consultation with the people. The Forest Service should be subservient to Indonesians' development, not that of "foreign enterprise."[27]

Dutch land management followed a different ideology, however, one that placed colonial and European corporate accumulation before the well-being of smallholders and other Indigenous peoples. Agrarian peoples' activities and rights were circumscribed and in some cases extinguished entirely. By 1930 plantation areas in West Sumatra had roughly doubled from 1900. These new plantations pushed autonomous smallholders farther up into the forest reserves, while a proletariat of petty traders and small producers expanded in the villages, towns, and cities. In response, the Dutch government authorized forest foremen to wield police powers in 1937.[28]

Criminalization of all rural work independent of the colonial commodity crop economy ensued. Forest foraging, hunting, and homesteading were to various extents all banned, alongside already illegal livelihoods like smuggling, scavenging, and thieving. Still, colonial control was not omnipresent. In the face of colonial efforts to control population movements, new dynamics of spontaneous migration developed, wherein the dispossessed established homesteads as smallholders in the patches of forest frontier that remained. Even with Dutch efforts to close them, these frontiers filled up. With luck these migrants could find an unclaimed plot to call their own or purchase, lease, or marry their way into a land claim.[29]

Over the decades colonial power was funneled into efforts at land control. The enclosure movement, while fraught at every step, served to establish the centerpiece of the Sumatran colonial project: industrial plantation expansion. No humans will be willing to do any exploitative work unless all other, relatively better livelihood options are inaccessible.[30] So colonial policy and management targeted the land frontiers—in this case the upland cloud forests—to try to reduce the ability of plantation laborers and smallholders to escape into other kinds of work.

With less than hegemonic but considerable resources, the Dutch applied the Agrarian Act. Spatial plans, maps, and surveys provided visualizations of the colonial doctrine of progress. Cartesian cartography gave colonial agents a vision of the future to act upon. It also allowed colonial authorities to map out resource use, agriculture patterns, and even peoples' living arrangements. Technologies of power established the lines of development. The Euclidean grid provided the coordinates that outlined colonial spatial expansion.

Land concession boundary lines marked the space within which colonial progress in the forests was achieved. These same markers enclosed and dispossessed smallholders, shutting them out from their agricultural lands and forests.

POST-COLONIAL CONTINUITIES

As a legal technology of dispossession, the land concession has endured. It remains the primary mode of state-corporate land dispossession in Indonesia, as it does in nearly all post-colonial nations. Land concession in hand, agribusiness and other resource extraction corporations have enacted regimes of violence and radical transformations of agrarian landscapes.[31]

In the 1960s, Indonesia's New Order planners returned to the colonial-era practice of creating new land concessions as a primary technique of dispossession. Dusting off colonial policy and European-scientific logic, the New Order unleashed a wave of dispossession across the archipelago. Coming along with the land concessions were ideas of forests and farms as a potential "natural resource" to be developed. Once again, peasants and Indigenous peoples became squatters on their own lands, criminalized.[32]

For the New Order's military leadership and administrators, Casiavera's existing livelihoods were nothing more than a barrier that prevented the unlocking of "development," "modernization," and "efficiency." As Steven Lansing observed in a detailed study of the New Order's agrarian disruptions in Bali, "The only question about the traditional social system [for Suharto's planners] was how much resistance it might offer to the spread of the new technologies."[33] Codified by the 1967 Basic Forestry Law, a state-corporate agribusiness complex empowered by military institutions came to control the land with a pervasive atmosphere of violence.[34]

It was a restructuring of territory on the largest of possible scales, in which both society and a breathlessly large biodiversity were hammered into a centralized politics of natural resource extraction. Woe, tragedy, and chaos joined the making of fortunes and new opportunities for extractive industries in the countryside. While the New Order claimed the sole authority to mediate between corporate interests and its citizens, the administration also claimed vast property rights vis-à-vis its rural populace. The regime provided monopolistic protections and resource access to foreign corporations, which then worked closely with the state's armed forces and private security to dispossess previous resource managers. An extractive industrial state emerged.

The state-corporate impulse toward control, of exerting power across the landscape in order to exclude those who pursue their livelihood there, reached its peak with the Agrarian Directorate of the Department of Home Affairs (led for its first thirty-two years by army generals). The directorate took up a policy of land dispossession to support the expansion of a range of industrial development projects. Starting in 1969, the directorate canceled hundreds of thousands of hectares of existing private land titles and began the de facto transfer of land control from smallholders and small-scale individual entrepreneurs to state-sponsored industry for infrastructure, natural resource exploitation, and agro-industry.[35]

It surprised no one that these logging concessions and plantation development sites were not the "empty" lands they were often portrayed to be in the sanitized New Order lexicon. Many sites belonged to local communities under customary claims. Industrial zones replaced smallholders' agroforests, vegetable fields, and fallows. The lion's share of new timber, mining, and plantation sites include documented cases of individuals and groups working to exert customary land claims and demanding compensation for lost forests, agroforests, and farmland. To the extent that corporations were able to develop these contested sites, it was because they were able to use harassment, intimidation, and violence against prior claimants, or they were able to make payments to entice existing customary owners to give up their claims.[36]

This authoritarian economic state allowed little room for dissent. From 1965 to 1966, the peasant and landless laborer unions and their supporting political parties were extinguished.[37]

Indonesia's many competing social organizations were redirected, under military agitation and widespread social violence, into the ideological concrete flume that was Suharto's New Order regime. Centralized management of the vast archipelago was achieved with the persuasive control of a "purged, streamlined and militarized" bureaucracy that, once solidified, was the dominant instrument for suppression of dissent.[38] Out in the agrarian landscapes, extractive industry operations became militarized enclaves. Company security guards and Indonesian mobile brigade police units (Brimob) guarded logging concessions and plantations, which maintained barracks, checkpoints, and watchtowers at key land concession access points.

Beyond the Aren volcano, the Agrarian Directorate allocated land concessions primarily in Sumatra and Kalimantan, where state and corporate authorities imposed new land uses with little regard for existing ecologies. From 1967 to 1975, fourteen logging concessions (Hak Pengusahaan Hutan) were given to foreign companies totaling nearly 3 million hectares, primarily in the Bukit Barisan. Seventy-two logging concessions were created as joint state-corporate ventures that covered an additional 7.6 million hectares.[39]

By 1982 the Agrarian Directorate had developed new spatial planning technologies with financial and technical support from the World Bank to transfer property rights of more than one and a half million hectares of land to corporations. Local government land acquisition committees constituted by New Order officials in Jakarta identified and recommended lands to be transferred to corporate users. The Ministry of Agriculture and Forests directed yet another wave of dispossession.[40]

The resulting expansion of agribusiness control of the land and resources was likely the largest wave of forcible land dispossession in Indonesian history. From 1992 to 1998 the Ministry of Forests and Agriculture created three million hectares of plantation land to be managed by twelve hundred foreign and national companies.[41]

In West Sumatra, from 1989 to 1990 the National Land Agency issued permits for fifty-five large-scale industrial land concessions in the province, mostly for oil palm (forty-one estates) but also for rubber, tea, and logging operations. These concessions encompassed more than three hundred thousand hectares of land in the province.[42] The governor and district heads (*bupati*), all appointed by the New Order, authorized these state land leases.

On the Aren volcano, this wave of dispossession began with the Dona cattle ranch then expanded across the Aren uplands in 1974 with a state-owned cattle ranch that dispossessed hundreds of smallholders on the eastern slope of the volcano. A decade later, in 1983, a twelve-thousand-hectare rubber plantation opened at the northern base of the mountain. Shortly thereafter a tea and ginger plantation opened as well. By 1989, when a Japanese investor announced plans for two more ginger plantations on the volcano, one at Dona, the local newspaper, *Singgalang Independen*, gave voice to growing discontent with the situation when it announced that the Aren volcano was being "ogled by many parties" because the government had "opened the door for investors."[43] According to the newspaper, the Regional Body for Planning and Development (Badan Perencana Pembangunan Daerah, BAPEDDA) was seeking still more investments in farming, agriculture, livestock, and fisheries and lecherous investors were targeting the volcano again, raising the threat of still more dispossession and exploitation that would replicate the already established Dona Company in Casiavera.

THE NEW ORDER'S PINE PLANTATION PROGRAM

Along one of the Aren volcano's ridges that defines the village of Casiavera's borders, just outside the boundaries of the Dona cattle ranch and plantation, is a mixed-broadleaf forest that became the site of a New Order timber planting project. Across Sumatra, the Ministry of Forestry planted similar pine plantations throughout the 1970s and 1980s. In Casiavera, the monocrop stands came to radically simplify the forest, a reduction from thousands of plant and wildlife species to a single species of Sumatran pine (*Pinus merkusii*).

In 1973 a regional newspaper reported the start of the Ministry of Forestry's pine plantation project with the planting of 150,000 pine trees on the volcano, along with plans to plant another twenty-five kilograms of seed in the near future. The planting was to be done on what they called "bald" or "barren" land (*tanah gundul*), in order to prevent erosion and the loss of the topsoil.[44] Sumatran pine trees only grow in degraded or rocky soils in full open sun. Where these exposed soils and grasslands exist in the Bukit Barisan, they typically mark the site of past use. One collection of practices created these exposed soils, relating to the ecological damage and abuse of both smallholder cultivation, like upland tobacco monocropping, and state policy, like Dutch colonial forced coffee cultivation and rubber plantations. A very different collection of practices created the patches of grassland, namely smallholders setting fires to encourage the growth of fire-adapted grasses for fodder to feed livestock.[45]

The pine program was the clearest instance on Aren of the New Order using environmental discourse to justify state intervention. State representatives portrayed smallholders as the cause of land degradation. The solution was given to be state management of the pine plantations as a form of "rehabilitation." Foresters joined corporate landholders to be especially critical of smallholders, who were labeled "squatters" who used "slash-and-burn" to clear vast swaths of "virgin" or "primary" forest that were of the upmost importance for the goals of environmental conservation and economic commerce.[46] Smallholders were "backward" and "destructive" agents of forest degradation.

The Ministries of Agriculture, Forests, and Transmigration propagated such myths and misrepresentations. The continuity from the colonial era is striking. In the 1920s and 1930s, when the Dutch were working to close the forests to smallholders, they too said they were acting in order to prevent erosion, deforestation, and watershed damage, most often spoken about as the "hydrological" problem. These environmental problems, the colonial line went, were best countered with state-enabled corporate control. Take the 1921 statement by the forester A. J. Beversluis that "irregular wood felling by local people" in West Sumatra should be combated with "increased regular exploitation by private enterprise."[47] Beversluis was the very same forester who had overseen the reforestation of the upper flanks of the Aren volcano with the fast-growing timber trees *Albizia, Dodonaea* and *Casuarina*, as well as the enforcement of this planted state forest's boundaries.[48] The commonality across the colonial state and the New Order ministries was the desire to transform changing, complex, humanized landscapes into fixed, easily legible locales of state control.

Environmental damage as a central development problematic was the New Order revival of European scientific forestry. Smallholders were constructed as invaders, out of place. Smallholder movements were labeled encroachers and taken to pose a threat to the function and purity of their locations of arrival. They were marginalized, criminalized, and pushed out of forest landscapes. They were denied the right to continue their own local landscape practices; their actions were depicted as "degrading" by experts of many kinds, whose own technoscientific knowledge was to be the solution.

In many ways, New Order views of smallholders' tendency to damage forests are the product of arguments put forward by Western scientists at the origins of tropical forest-farmer scholarship. Well into the second half of the twentieth century, a belief in the inherently damaging role of forest agriculturalists was the dominant view. Published the same year the New Order began, J. E Spencer's *Shifting Cultivation in South East Asia*, emerged as the definitive account of tropical forest agriculture as ecological destruction.[49] His careful attention to the social and ecological processes of clearing plots of forest for planting notwithstanding, Spencer nevertheless made the mistake of assuming humans' influence on the landscape to be inherently negative.

The use of the fallow and tree planting—that is, swidden fallow agroforestry—was leading to the "destruction and waste" of forest resources and a problematic altering of "wild landscapes." Focusing on the fallow, Spencer had no appreciation for the forest cultivation that followed clearing and fallowing, leading him to believe that rainforest agriculturalists had little influence on the composition of the forests other than their destruction.[50]

The New Order seized on similar ideas of smallholders being an environmentally negative force to justify taking control of Aren's ridge forests and plant pine. Each regional Ministry of Forestry office included a security division composed of police, army members, and foresters that worked to keep smallholders ("encroachers") out of the forests. Arrests for illegal logging in the pine plantations were directed against the few smallholders who were sufficiently brave or desperate (or some combination of the two) to cut timber in the plantations without official or illegal sponsorship by someone more powerful than themselves.[51]

The "land rehabilitation" pine planting program became a social lightning rod at the center of state dispossession and corruption. Rogue officials (*oknum*) used the program to claim agricultural forests as "degraded" land, log all timber of value, and, only sometimes, replant the land in pine. Whatever pine was planted became their own, sold illegally without tax payments. Unscrupulous Ministry of Forestry officials often worked with police and military at the center of the logging gangs.[52]

With the fall of Suharto in 1998, critics became more willing to speak out about the contradictions between state destruction of agrarian environments and discourses about damaging smallholders. One outlet for these criticisms was the Radix email listserv, a digital platform for anonymously sharing uncensored information during the late New Order and into the first decade of Indonesia's following democratic period. In those years a palpable outrage pulsed through the listserv, stemming from the New Order's widespread destruction of Indonesia's forests and subsequent misplaced blame on smallholders and Indigenous peoples. One critic was "very disappointed" because the government refused to answer the critic's letters about the state's contribution to the death of the forests and the "bastard forest destroyers' use of [smallholder] scapegoats."[53]

A second critic used the listserv as an outlet to elaborate the biting truth about the New Order's agrarian transformations, writing to the Radix audience with a nihilistic critique:

> Almost 75% of the State Forest is severely damaged and ill again deserved to be called forest. Most of the reforestation program is only a theory and slogan . . . the pine trees that are planted are no more than the remains of all the corruption. . . . I guess all agree with this. The forest damage is the same in all other parts of the country.

The Forest Pig Vandal-Thieves, including concessionaires, made an ill-kept promise of reforestation, and the corresponding program has only been used as a pretext to a "Forest Community Dialogue" that is destroying our forests. The dominant Suharto-esque Department says forest encroachers are the problem to fool people into thinking that cultivators are pests. The end result of forest destruction is illegible to most people. So, the corrupt nature of officials and soldiers who led most forest programs have brought the current crisis of corrupt behavior in the reforestation program, which has seen less than ten percent of planned reforestation realized nationally.[54]

Still uncertain about the limits of the post–New Order's newfound freedoms, this critic chose to identify himself or herself only as Kopassus, the name of Indonesia's special forces army division. "Kopassus" points out the irony of New Order officials' appeals to increase vigilance against timber theft in the State Forest even while members of their own ranks looted the forests. Most interesting, though, is the way that the critic rejects the New Order's negative depiction of forest farmers as nothing more than a pretext for officials' acts of thievery.

One of the Ministry of Forestry officials responsible for implementing the pine plantation program in West Sumatra was Irfan. Retired, he moved to his wife's hometown on Aren to live underneath one of the volcano's ridges planted in pine. In 2015 he took me for a drive in his green Isuzu Panther SUV to see the remaining pine plantations above Casiavera and share his thoughts on the program. The "land rehabilitation" program began in 1966, using pine planting for "reforestation," "re-greening," and "conservation," Irfan told me. He was much more cautious than the critic "Kopassus," but his unease with the social strife in pine plantations was noticeable.

Irfan told me that the program made all the lands where pine was planted state-controlled land, although he said he worked to allow a certain number of villagers to harvest timber for their own profit. According to him, the program was designed to maintain soil fertility on already barren lands, but it ended up being implemented in a different way.

Back at his home at the foot of the volcano, Irfan walked me to a wall in his living room, where along with framed photographs of his children's college graduations and marriages hung a letter signed by none other than

Major General Suharto, from 1991, thanking Irfan for his contributions to the Land Rehabilitation project. Still loyal, Irfan did not want to talk about the way the pine plantations had transformed land control or the "squatters" who cut and burned the pine plantations. He only mentioned that the long-running program was not put into practice according to its original design.

The pine monocultures planted across the ridges above Casiavera were a result of the New Order's reordering of the forests. They grew in neat rows, all the same size and age: the New Order monoculture of the mind made material on the landscape. And many people in Casiavera took the pine plantations' persistence as an insult.

I learned as much when I took my first walk alone, without Irfan, along the ridges to see the pine plantations up close. This time a man greeted me during my walk along a trail through the pine. Introducing himself as Ramli, he told me that this ridge above Casiavera was "sentenced to become pine" during the New Order. He continued, "We did not agree. And we still do not agree." Ramli's eyes were shaded by his faded ball cap, but I could hear the emotion in his voice. "The people here do not forget. It was during the 1970s. We watched as they cut all the trees and then planted the pine. We were scared. We did not do anything. I was just a youngster in grade school."[55]

On a handful of subsequent walks through the pine forests I learned of the difficulties of pine monoculture for smallholders. It was difficult to grow anything between the trees, which were planted very close together, creating a deep shade. Their acidic needles leached into the soil, making it harder to grow other crops. The community was also tired of the ambiguity, the de facto illegality of their unregulated cutting and selling of the timber. Their discontent took material form in 2015, when they burned down a part of the pine plantation. "We were scared back then. We are not anymore. Now, this is really the people's forest (hutan rakyat)." Unlike the heavily policed industrial palm oil and timber plantations, the New Order pine forests are now mostly unregulated. Ramli told me, "Occasionally, someone from the Ministry of Forestry comes and says we need to stop. But nothing happens to us. It is not clear who owns them. And this is our community. Where are the state's [ownership] letters?"

A handful of sawmills would accept pine logs on Aren. The boldest reclaimers took the tallest, straightest pines to sell for building houses. There were also a handful of stores at the base of the volcano selling the

FIGURE 6. Reclaiming within the pine plantation monocultures.

timber. Along with the felling, fires in the pines were common as well. It is impossible to know how many fires were political acts of arson, but many suspected the bulk of them were.[56]

Reclaimers made use of the fires' boost of the nutrients into the soil to establish their own fields. After opening up the pine forests, smallholders began intercropping gardens of chili, corn, and grass for their cows. They planted cassava and banana. Ramli planted mahogany, durian, cacao, lime, and grass in a clearing within the pines. Pine stumps were scattered throughout his plantings. The plantations remained mostly quiet, though. There was not much foot traffic. Much of the land was still under pine.

The pines standing along the ridges above Aren were a reminder of the exclusion of smallholders from the forests and the corresponding reduction of autonomy that accompanied the arrival of New Order Forest Police and their surveillance and persecutions in the not-too-distant past. The pines marked the decades-long authoritarian control of the landscape, a control that created plantations of rubber, oil palm, and timber that took a particularly militaristic and threatening monocrop form, a form that workers from Casiavera worked to undo, intent on reestablishing their smallholder forests.

CHAPTER 3

The Plantation Lifeworld

By the 1970s the dispossession of Casiavera's smallholders was complete. The Dona cattle ranch was in full swing. Overseers managed teams of laborers to tend the livestock and their feedstocks. Almost certainly unknown to the worker and overseers both was the gyro-stabilized satellite camera orbiting over the volcano taking photographs of the plantation every few months. My account of the plantation lifeworld opens here, not with the land but with this satellite camera orbiting above it, a satellite that was part of the Central Intelligence Agency's (CIA's) covert Corona surveillance program, the world's first spacecraft to autonomously create photographs of the Earth's surface. That these photographs came to exist at all speaks to the way that Casiavera's smallholders' dispossession was determined through the aligned political practices of the CIA and the New Order that extended new forms of spatial and social control across the landscape itself.

All throughout the loss of Casiavera's collective land, the destruction of the smallholder landscape, and the creation of the Dona cattle ranch, Corona provided photographs of Casiavera to CIA analysts in their Foggy Bottom offices in Washington, D.C., the same analysts who played an instrumental role in creating the conditions for Major General Suharto's rise to power and the formation of Indonesia's New Order regime.

The Corona project began in 1958, directly following the failure of a short-lived, CIA-funded rebellion in Sumatra that was, as CIA director Allen Dulles described it to the National Security Council, part of the US government's efforts to "overthrow communist totalitarian regimes."[1] More accurately, the CIA's covert war in Sumatra's Bukit Barisan uplands was a reactionary intervention to stop the newly post-colonial nation from moving toward a non-aligned, syncretic geopolitics that rejected international capitalism and embraced the republic's influential Islamist political parties as well as the Indonesian Communist Party (Partai Komunis Indonesia), then the largest communist party outside of China and the Soviet Union.

With CIA funding, training, and weapons, the Sumatra rebellion and its affiliated Revolutionary Government of the Republic of Indonesia (Pemerintah Revolusioner Republik Indonesia, PRRI) sparked an Indonesian constitutional crisis and engulfed West Sumatra, including Casiavera, in war. But the pro-plantation, pro-capitalist rebellion did not have any real popular support. The freedom fighter turned first president of the republic, Sukarno, labeled the PRRI a "separatist gang" (*gerombolan separatis*) and directed the Republic Army to act quickly to contain PRRI forces in Sumatra's Bukit Barisan mountain forests.[2] Facing an impossible situation, shortly thereafter the rebellion's military leaders emerged out of the forests to sign their surrender in the upland urban center of the Minangkabau territories, Bukit Tinggi.[3] The CIA's assistant director of national estimates saw this quick defeat, bereft of grassroots support, as a "geopolitical disaster" that became the "greatest propaganda and gain" for the Communist bloc.[4]

On April 8, 1958, CIA director Allen Dulles presided over a room of the chiefs and directors of intelligence from all the branches of the armed forces, as well as the Federal Bureau of Investigation. The men gathered in a session of the Intelligence Advisory Committee to mull over the ramifications of PRRI's easy defeat for other US anti-communist wars in Asia. The committee saw lack of intelligence as key to the failure. Its proposed solution was as surprising as it was transformative: the Corona reconnaissance program. In the April 8 meeting, directly following their discussion of the PRRI debacle, the group received a presentation by one General Schow of the Critical Problems Collection Committee for a new kind of covert intervention in social control, "a high priority Air Force project for the development of an earth satellite reconnaissance system which was already underway"—that is, Corona.

After their failure in Sumatra's Bukit Barisan mountains, the North American soldiers and spies realized they had badly miscalculated Sumatra's political forests. They recognized they needed to make the world's rural spaces more legible to their analysts, especially the nearly impenetrable

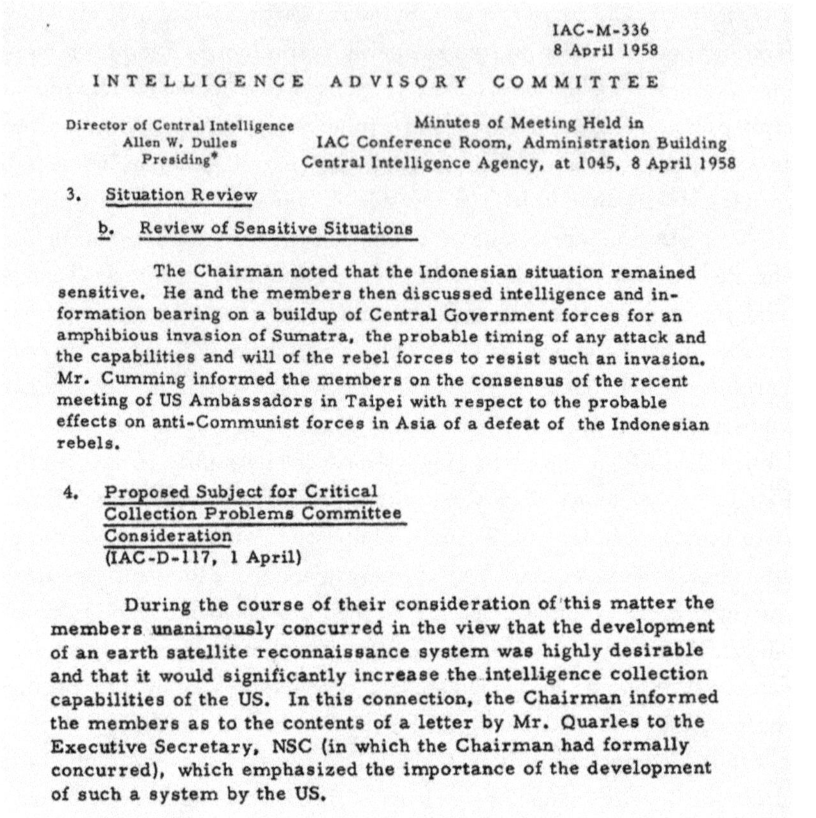

IAC-M-336
8 April 1958

INTELLIGENCE ADVISORY COMMITTEE

Director of Central Intelligence
Allen W. Dulles
Presiding*

Minutes of Meeting Held in
IAC Conference Room, Administration Building
Central Intelligence Agency, at 1045, 8 April 1958

3. Situation Review

b. Review of Sensitive Situations

The Chairman noted that the Indonesian situation remained sensitive. He and the members then discussed intelligence and information bearing on a buildup of Central Government forces for an amphibious invasion of Sumatra, the probable timing of any attack and the capabilities and will of the rebel forces to resist such an invasion. Mr. Cumming informed the members on the consensus of the recent meeting of US Ambassadors in Taipei with respect to the probable effects on anti-Communist forces in Asia of a defeat of the Indonesian rebels.

4. Proposed Subject for Critical
Collection Problems Committee
Consideration
(IAC-D-117, 1 April)

During the course of their consideration of this matter the members unanimously concurred in the view that the development of an earth satellite reconnaissance system was highly desirable and that it would significantly increase the intelligence collection capabilities of the US. In this connection, the Chairman informed the members as to the contents of a letter by Mr. Quarles to the Executive Secretary, NSC (in which the Chairman had formally concurred), which emphasized the importance of the development of such a system by the US.

FIGURE 7. Intelligence Advisory Committee meeting notes, April 8, 1958.

landscape behind the communist Iron Curtain. More than anything, Corona was an attempt by the US state and military to build a grid of control of the planet.

As the Corona satellite camera passed above Casiavera in what military aerospace engineers would come to call the "reconnaissance orbit," for the first time CIA analysts felt that they had engineered more than a map of the landscape; they engineered a map of the social terrain. Arthur Lundahl, head of the Corona National Photographic Interpretation Center, claimed that with Corona photographs he could accurately determine where people were born and buried; their religion and level of formal education; and what they ate, wore, and manufactured.[5]

As a realization of the surveillance panopticon at the scale of the Earth, Corona was indeed able to visualize the world with an astonishing clarity and detail. However, it ultimately led the CIA to err. The landscape was revealed, but not all aspects, as CIA leadership believed. When CIA analysts passed thousands of photographs of Indonesia and other parts of Southeast Asia across their negative interpretation machines, they saw mostly clouds. When their orbiting cameras did capture a clear view of Indonesia's agrarian environments, the analysts saw mostly the tones and patterning of wet-rice terraces, smallholder farms, and forested uplands.

For the analysts, the sum of the details visualized with Corona was but an undifferentiated mass of rice paddies and forests blanketing the landscape. So it went across the whole of Southeast Asia. Indonesian, Vietnamese, Burmese, Cambodian, and Laotian rural peoples devoted themselves to livelihoods and mobilizations that did not create large-scale human constructions easily spotted from space. Corona's analysts were almost entirely unable to report significant ground military activity across Asia, and certainly not the political thoughts of the people who lived in it. Smallholder ways of life, the terrain, clouds, and rainforest canopy provided cover against Corona.

The CIA's undifferentiated perception of the Southeast Asian countryside—derived from their imperfect Corona visualization methodology—fed the West's paranoid fears of peasants as a vast, unified, and destructive communist force. Even while spatial technology informed political-military strategy, it contributed to an ungrounded and misleading empirical knowledge that eventually brought into being nothing less than the Cold War–era Southeast Asian killing fields of Vietnam, Cambodia, Laos, and Indonesia. Surrounded by spatial information but with little understanding of Southeast Asian societies, the United States engaged in terrible, ill-informed geopolitical interventions. The subsequent wars brought smallholder political setback, violence, and suffering on the scale of entire societies. With US support, in Indonesia the New Order's pro-West, capitalist military dictatorship would eventually completely extinguish the peasant movements on Aren and across the archipelago.

Corona satellite photographs exist as artifacts of this tragic history. From May 1962 until the end of the project in 1976, Corona satellites produced nearly one hundred images of the Aren volcano. These photographs were created with the masters' tools, for the state to control social and ecological

1964   1976

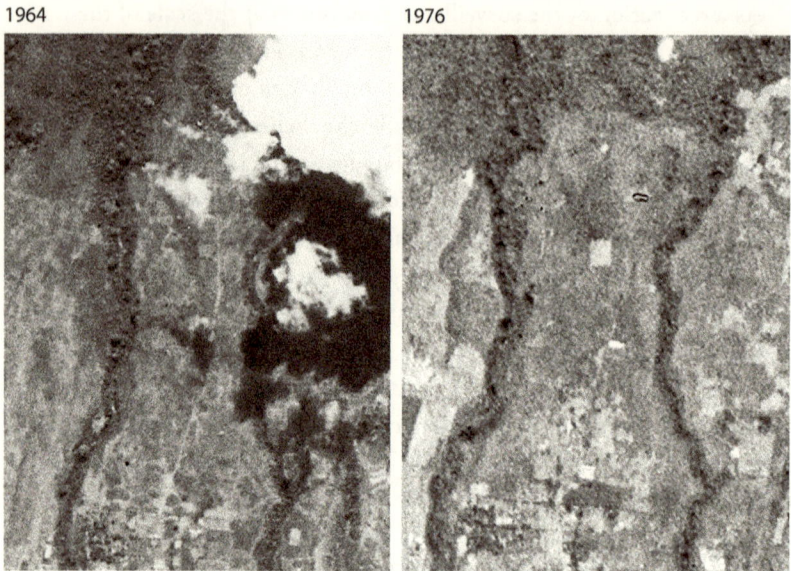

FIGURE 8. Corona satellite photographs illustrating the New Order dispossession of Casiavera's smallholders for a corporate ranch and plantation, comparing 1964 and 1976.

life. Yet as a visualization of the changes that unfolded in Casiavera, it is possible to repurpose these satellite photographs for a more liberatory purpose, one that documents the landscape simplification and destruction that the Dona Company wrought above Casiavera.

My reading of the Corona photographs begins with the clearest view of Casiavera's land before the Dona ranch was established, on December 22, 1964. A contextualized view of this 1964 image shows a mosaic of rice terraces, vegetable gardens, and forests. Networks of footpaths connected small mixed plots of vegetable and tree crops. The lower parts of the land were more intensively planted. The upper reaches transitioned to the cloud forest above, on the steepest slopes of the volcano's upper reaches. This was a smallholder landscape, cultivated in heterogenous, small-scale patches.

Such forest-field mosaics lack defined boundaries and contain diverse ecologies. They are complex peasant landscapes that have greater flux and change across the micro and landscape scales than the monoculture plantations that have often replaced them. It was precisely this flux and change that the plantation was designed to simplify and destroy.

Abril, a man born in Casiavera who worked most of his life on the ranch and plantation, first as a laborer, then as a manager, and then as village head of Casiavera in the late New Order, remembered how after Japan's withdrawal from the island and the declaration of the independent Indonesian Republic, Casiavera's customary leadership re-formed their control of the land.[6]

Abdul, once head of the Casiavera Community Council, wrote as part of a history of the land that "by 1951 our people used and managed the land as an agricultural field (*dikelola dan memanfaatkan untuk bidang pertanian*)."[7] A second man, an elderly family-lineage leader who has since passed away, wrote that he personally worked the land, planting a mix of crops, starting in 1945, joined by many others. The war and independence provided an opening, and the community seized it to make the land part of the village's lands once again. They used the land as village collective land (*tanah ulayat nagari*), a resource thought of as owned in collective by all the residents and perceived members of the *nagari* community.

Some families planted sugar palms to tap for their valuable sap. Others chose to plant sugar cane and vegetables. A few planted tree crops like cloves. Still others grazed their livestock on the highest reaches of the land below the steep cloud forests. All of these smallholders lived down below on their family lineages' lands, where they worked rice paddies, home gardens, and fields.

Back then, Abril told me, no one lost homes or rice paddies to the company's dispossession of the land, making it less of a threat to the daily survival of the community as a whole than it could have been. What's more, many families in Casiavera had members who worked trading in the markets or as entrepreneurs, teachers, and bureaucrats, further reducing their dependence on the land that was lost.[8] Still, the dispossession ruptured families' access to the collective land, a valuable economic resource. The company also refused to let the community pass through the concession, making access to the cloud forests above the concession much more difficult.

Even beyond its material effects, which were considerable, the dispossession was a challenge to the *nagari* community's sovereignty. Coming as it did at the end of the Indonesian Republic, the dispossession was a clear indication of the eroding political position of Aren's smallholders. As it went across Sumatra, the rise of the land concession system and the industrial mining, logging, and agriculture that depended upon the land concessions was foreclosing the promises of independence.

By 1976, eight years after the New Order's transfer of control to the Dona Company, a second Corona photograph shows land use to be radically transformed. Company offices and barracks are seen at the center of the concession area. Grass for cattle feed grows where once small patches of diverse smallholder agroforests stood. The network of footpaths smallholders used is gone. Much of the lower portion of the land is exposed soils, indicating that it has been degraded through overgrazing. On walks across the collective land with Abril, he remembered back to his youth, telling me that when Dona began operations, laborers from the upland plain below the volcano came and burned the land to clear it. Dona's laborers built a workers' barracks, offices, and cow and horse sheds. Teuling brought in an overseer and technicians from Java. Company laborers planted elephant grass and other livestock feed. The patches of forest, the community's vegetable plots, and sugar cane crops all became cattle fields. "There was not a single tree [that remained]," Abril tells me.

With their land concession in hand, company managers banned peasant use of the land. The company allowed Casiavera's smallholders to harvest their last crop of annuals, then told them to leave their land. Dona's overseers marked their control of the land by building a fence at the lower boundary of the land and a road up to its upper boundary. They began their work in the center of the concession, burning it and then planting grasses for cattle grazing. Subsequent burnings of any forest regrowth determined the emergence of a grass-dominated fallow. The cycle of grazing and burning destroyed the existing forest-species seed bank. The microclimate of the pasturelands prevented any germination of forest species brought in by wind, bird, or human seed dispersal. This anthropogenic pasture provided fodder for the livestock, allowed to range free across the land. As the landscape was transformed it allowed ever-fewer species of life to live upon it. Corporate managers replaced the hundreds of species that had thrived in the patchy, mixed fields and their fuzzy transition into the rainforests above with cattle pasture that was created to sustain only two life forms: cows and the grass they eat.

NEW ORDER CONTROL AND FEAR

Zain grew up under the New Order to become one of the most well-educated people on the volcano. A quick talker, he worked as a lawyer for the local government. We often hung out on the weekends, when he would wear motor sports jerseys covered in Michelin, Yamaha, Kawasaki, and Indonesia Off Road logos. Perhaps it was his legal training that made him one of the most eager to discuss life on the volcano under Suharto. He summed up life under New Order dispossession simply:

> During the New Order we were afraid. They accused us of being part of the forbidden party (*partai terlarang*). And this was after the clash [the genocide]. The people could not speak![9]

Control and fear were interconnected and omnipresent. Even fifty years after the terrible genocide, Zain, a well-established, middle-aged man with no clear radical engagements, still couldn't bring himself to say the word *communist*, the label that New Order officials applied to his family and neighbors when he was a teenager.

It is difficult to convey the saturation of fear in society that the New Order instilled across the archipelago. One view into the tenor of New Order dispossession is given by the Sawito Letters, a series of documents that a collective of disillusioned elites penned in September 1976. Among these was the well-known West Sumatran Minangkabau statesman and first vice president of the Indonesian Republic, Mohammad Hatta. The founder of the Indonesian police and popular Javanese mystic Sukanto Tjokrodiat-modjo and a retired Armed Forces chief of staff, General T. B. Simatupang, also joined the signees. In calling for Suharto's immediate resignation the collective wrote:

> The people are enveloped by a fear of politics and the might of the authorities.
> The existence of indefinite detention without trial and other psychological threats intimidates people everywhere.
> Accusations of being remnants of G/30/S/PKI [communists], the Old Order, subversive, anti-development, a sower of hatred, a danger to political stability and national security are like the sharpest samurai swords and greatly feared by the people.[10]

The Sawito Letters painted the early New Order as an era of oppression, violence, and fear. Most sinister was the continued labeling and repression of anyone associated with leftist organizing and "anti-development" thinking, that is, anyone opposed to the state's control of the land and agrarian territories.

Yono, a resident of Casiavera who did not participate in Casiavera's reclaiming movement, told me that the kind of repression and fear that the

Sawito Letters denounced occurred in Casiavera. According to Yono, the state labeled Casiavera as a "Gestapu" village to justify Dona's 1968 land lease. Doing so allowed the state to completely erase the people's right to self-determination: "The [land] permit came straight from [the national government] in Jakarta. They did not even ask permission from our Council."[11] Yono was matter of fact about the injustice and without any outward expression of fear. It turned out that Yono's courage to talk with me stemmed from his position in society. In the early days of the New Order, Yono had joined a paramilitary squad tasked to arrest suspected communists. He went on to be a New Order Forest Police officer. "I carried a gun," he told me with pride as he tapped his chest. Had he been part of the opposition to the authoritarian regime, no doubt his confidence, and willingness to talk with me, would have been much reduced.

Many others told me about their fear when Mahmud Teuling and Dona arrived to take control of the collective land: "We were afraid"; "Teuling's permission came straight from the capital. We were afraid of him"; "They were arresting communists in this village!"[12] Still others chose to not recall anything at all. Alone with Oih out among his banana palms, the reclaimer told me, "The history is that Dona got the land because we were all communists. But I don't know anything about that, it could be correct. It could not be."[13] Agoez, an older, skinny, gentle-spoken reclaimer, similarly told me, "I don't remember [about when Dona began operating]. My mind is not so good." Yet Agoez worked for the Dona Company for seventeen years, and he was willing to tell me that Mahmud Teuling had a love of racing horses.[14] Agoez's selective memory is in fact a deeply ingrained strategy of survival. Agoez and the others lived through a time when discussion of specific topics could be enough to bring condemnation, jail, exile, or death.

Casiavera's survivors of the 1960s killing, incarceration, criminalization, and land dispossession almost exclusively choose to not talk about the violence. Talking about dispossession was difficult because of the many layers of trauma it entailed. These memories and stories were guarded, held close in attempts to reestablish a sense of normalcy following traumatic events and avoid attracting the perpetrators' attention.

As my conversations deepened in Casiavera, I came to be more aware of the silences between what men and women told me they remembered. Previously chatty, lucid folks would jump to what became for me familiar refrains: "I don't remember anything about that." "My memory is not too

THE PLANTATION LIFEWORLD   67

good." Or, more simply, "I do not know anything about that." The way these silences were given to me made me believe they were strategic. Such strategic silences support survival and self-preservation. They suggest it is smarter to stay silent than it is to share one's dangerous memories with a visiting anthropologist.[15]

Agents of New Order violence persist in positions of power today. These men (and they were almost exclusively men) do not feel the need for strategic silences in the way that their victims do, or like those associated with leftist politics, intellectualism, and art. My experience in Casiavera mirrors an important ethnographic reference point of the early New Order, Joshua Oppenheimer's 2012 documentary film *The Act of Killing*. In it, one of the New Order's executioners, Anwar Congo, reenacts for the camera the violent memories that haunt him. Congo spins his retelling of true stories of his murder of leftists while comfortably protected by Sumatra's ruling elite from reprisals and retribution. As an aggressor, Congo does not fear breaking his silence; he only fears his own violent memories, which torment him without end.

Similarly, the members of the community most willing to talk to me about this troubled political history were New Order officials, including a retired policeman, one of the village heads of Casiavera under Suharto, and Yono, the retired forest policeman. All these men had participated in a thirty-year history of authoritarian violence. That they remain respected and powerful speaks to the continuities of New Order violence and fear in the Bukit Barisan.

Fear spreads quickly, but it does not dissipate with an equivalent speed. Fears of the New Order remain lodged in Casiavera's collective memory. Every time politically motivated violence or repression reappeared, these fears were triggered yet again. One night a group of young men and I watched a news report on the murder of an anti-mining activist in Java. As the coverage jumped to interviews with eyewitnesses of the fatal daytime beating, a smallholder farmer who had joined peasant protests told us, "They do not have to hunt all of us, just two or three and everyone feels it. We all are silent."

Weeks later, I was not surprised when the fiftieth anniversary of the New Order genocide passed and no one in Casiavera's coffee shops remarked on the event to me. I was reminded of the date only by the front-page article in that day's newspaper. The story mentioned that the Jakarta chief of police had joined with the head of the Islamic Defenders Front to use the fiftieth anniversary as a reason to hold a press conference to denounce "all communists everywhere" and honor the six Indonesian army generals and one lieutenant who were killed during the 1965 coup d'état that saw Suharto gain power. A few days later President Joko Widodo announced on live television that he would not be able to apologize to the victims of the genocide. It was strange, this apology for not apologizing. The *Padang Ekspres* ran a front-page photograph of a few young men burning a red hammer-and-sickle flag in celebration of the president's refusal.

The fiftieth anniversary of the genocide mixed with the strategic silences in Casiavera to bring fear not only to others, but to me as well. Fear is contagious. By the time news reached Casiavera of the murders of a smallholder in Jambi by the private security forces of a major agribusiness and an agrarian activist in Jakarta, whose book on corruption in the oil palm industry was going to press when he was stabbed in a Jakarta nightclub, I had learned enough to expect my own silence and fear.

In my home at the end of the road on the volcano, my thoughts began to return to the old critiques of the oil palm industry I had written and my belief that the government was eager for me to not continue them. I thought about who else pays attention to critiques of the plantations, and what these people's typical methods of instilling fear and silence are. I began to avoid trips to town, to keep a lower profile. For a period, I stopped conducting interviews with government staff, worried about who they might report to. And with much regret, I could not help but wonder which of the old men on Aren who had welcomed me to their homes were, a generation ago, among the young men who took up knives to join the New Order's pogrom.

Power is necessarily local. It emerges and is expressed through interactions between people. On Aren, relations of power had a violent history. Everyday uses of violence of the kind that existed under the New Order have mostly receded into the past in Casiavera. It is not a particularly violent place. I did not hear any stories of thugs and toughs with the power to run daily life in the village, and there were no recent accounts of systemic state violence there either. But fears of Casiavera's violent past remained, along with the strategic silences these fears inspire. These fears and silences formed barriers to resistance and reform, for speaking out, working together, and changing the circumstances of life.

Although no one in Casiavera told me this, it is likely that lingering fear accounts for why only a minority of smallholders in Casiavera joined the

Indonesian Peasant Union. Consider what Arbi, a middle-aged photographer and naturalist keen on long walks across the volcano, told me about how he felt about his history: "I have fear, more in the past than it is now, but it is there. It creates boundaries, where I just need to be safe. It makes me stuck where there are conflicts, even though I want to do something, anything."[16]

The full depths of such fears across Casiavera remain uncertain, because the silences that fear instills in us make analyzing social repression difficult. Much more certain was the fear people in Casiavera had of the modern-day capitalist agrarian economy. As one Casiavera mother of two young men told me only half-jokingly, if her children decided to leave to look for work in the city, the only thing they would find there is a mafioso, who would sell them into a forced labor contract.[17] We all chuckled for a moment and then fell silent, the sons' expressions heavy with the risks of their still undetermined futures, before she changed the topic.

COOLIE WORK

The New Order dispossession of Casiavera's workers radically simplified not only the landscape but the social system as well. The agroecological knowledge that came from smallholders' relations with the land was interrupted. These relations took form when the Japanese occupation of Sumatra during World War II came to an end. The Japanese withdrawal left Casiavera's smallholders to use the land as they saw fit for the first time in half a century. Fifty or so of Casiavera's families started planting sugar cane, chilis, and sugar palms across the land.[18] Not more than a decade later, the Dona cattle ranch destroyed the peasant landscape. This time it was not a colonial power of dispossession but an authoritarian military dictatorship that made coolie labor the dominant relation on the land.

Mahmud Teuling, the owner of the Dona cattle ranch, built on his West Sumatran heritage, his position as part of the armed forces elite, and his accumulated capital to recruit powerful Casiavera residents, including some of the family-lineage leaders who served on the Casiavera Council, to support his cattle ranch and tobacco cultivation business. Teuling further consolidated his standing when he built a mosque below the land and made improvements to the dirt road connecting the village to the upland plain below.

Over time, Teuling found willing overseers from Casiavera, like Bajar and Abril, both of whom would go on to become village heads in the 1980s. They told me they worked for Dona because they "needed to live (*butuh hidup*)."[19] Abril was the head of a Dona labor team. He was paid a weekly salary to "control" the laborers and the land.[20] He told me that at first all the workers on the ranch were young men from Casiavera whose families had once used the land themselves. Almost two decades later, when the company began planting a ginger plantation, hundreds more workers joined up, including many more women and middle-aged folks. Almost all of them were from Casiavera as well.

Others chose not to work at Dona. Zed, a man of similar age to Abril, saw elements of exploitation in Mahmud Teuling's rule: "The people were taking wages while working on their own land. There was no justice (*keadilan*)."[21] Zed's rejection of coolie work was a real option for him, because he had another livelihood opportunity that did not depend on Casiavera's collective land: he was a New Order policeman. Another man born in Casiavera, Agoez, did not have any land, education, or money to buy his way into the New Order bureaucracy like Zed. He spent life as a laborer on Dona, where he "searched for a life (*cari hidup*)." He started working there as a fifteen-year-old. The land was covered in grass for the livestock. For him, this grassland landscape was "completely bare." There were hundreds of cows, more than one hundred horses, and some thirty workers. At first he was paid the "bachelor salary" (*bujangan gaji*)—1,500 rupiah a day, an amount equivalent to about one US dollar today. Then it was 2,500 rupiah, when he and his wife had children. He took care of the cow sheds, cleaning out manure and hosing down the corralled animals. It was poorly paid, menial labor. Of his time working on Dona, Agoez tells me: "I did not prosper (*Saya tidak ada berharta*). Because we did not have land."[22]

THE FALTERING PLANTATION

In the late 1980s Mahmud Teuling's fortunes declined. His health began to fail, and he lost control of his business dealings. His Kansas Cigarettes trading business had incurred large debts to Javanese tobacco traders. Then he lost a lawsuit with one of his biggest competitors, a Chinese Indonesian family business based in Java. To meet his obligations Teuling sold his herd at Dona and stopped paying workers. He was bankrupt. "He sold everything, until it was all gone," Agoez, the longtime company laborer, told me.

Not for the first time, and if history is any indication, not for the last, finance capital disconnected from Aren. The land went fallow, unused because of locals' fear of defying the state-sanctioned land concession boundaries. Teuling died a few years later, leaving a small inheritance to his children that included ownership of the Dona Corporation. His daughter, Monica Jessica Teuling, took possession of the Dona land concession and began overseeing day-to-day operations.

This second-generation Teuling went to work to convert the land into profit. Born in 1952 two hours' drive from Aren, Monica Jessica came of age with the New Order. After completing her studies abroad, she took her first job as an account manager at Citibank Jakarta. After a few years, in 1977, she moved to the World Bank's accounting division, where she would work for nearly twenty years.[23] In 1992 Monica Jessica Teuling began converting the cattle ranch into a tobacco and ginger monoculture, for export to Malaysia, India, and China.

With capital from a Javanese agribusiness investor, Dona transformed the land yet again. Supervising hundreds of people as day laborers, an overseer directed multiple rounds of tilling of the grazing land to prepare it for planting. Tractors arrived to turn over the soil, along with a fleet of trucks to transport the seed, fertilizers, pesticides, and herbicides. A few men from Java operated the heavy equipment and oversaw the work crews. Bajar and Abril organized the laborers into groups of ten workers each. Together with his son, who was just out of primary school, Agoez worked to prepare the land for the first ginger crop. "We planted the ginger, fertilized, sprayed pesticides, and harvested it." New land boundaries and blocks defined the long rows of a single monoculture crop. Dona's staff applied alienated labor to plant a single cultivar on dispossessed land—the defining processes of industrial monoculture

Daud, the manager of Casiavera's mosque and a peasant union member, was a young man when the Dona cattle ranch began operating. He grew up in his grandmother's home not more than a few hundred feet from the lower boundary of the land concession. Rather than seek work among the fifty or so mostly young, local men at Dona, Daud chose to migrate to Malaysia, looking for work in the cities. He returned home just as Monica Teuling began the ginger plantation operation on the land. Along with more than one hundred other workers, almost all from Casiavera, Daud signed up to labor on the plantation. Overseers managed the labor teams, along with the bulldozers, tractors, and trucks required to keep the operation running. Daud used a pesticide backpack sprayer for two months for the equivalent of five dollars a day before he left to find a different path.[24]

After tending Mahmud Teuling's livestock, Agoez recognized how much more difficult laboring in the ginger monoculture was. The pay was the same, but "the work with the ginger was so heavy. Taking care of the cattle was not that hard."[25] When the crop came to harvest, they spent their days collecting ginger from the soil, bending over at the waist with the half-hoe, a tool Cesar Chavez worked to ban in California because of the damage it inflicts on laborers' bodies three decades before Agoez spent his days swinging one. Wangga joined other women to work the pesticide and herbicide sprayers, but she left after two months because the pay, about 25,000 rupiah per day (equivalent to two US dollars in today's currency), was too low and she did not have time for anything else.[26] Like many, Wangga's family did not have any land to give or buy for her where she could start a homestead. After her short time with the pesticide sprayer, she knew she didn't want a life as a plantation laborer. So, like hundreds of young people from Casiavera at the time, she left. Wangga chose to move to the coast hundreds of miles to the West, where she could stay with family while she looked for work.

For those workers who stayed as the Dona Company turned the land into a ginger monoculture, company overseers bound them into a new relation of production. Abril, as village head, worked to ensure that dissent in Casiavera did not disrupt Dona's operations: "I had a job, I had money in my pocket. That was all that was important. Monica had the agreement, the land concession lease from the government, what was I to do?" Through the early 1990s Abril "let Teuling do her business. I respected the law and the plantation boundaries."

With plantation control established, coolie labor became the dominant relation for a good portion of Casiavera's residents as they did stints as laborers on the ginger plantation. They took up the familiar work of monocultures: driving tractors to till the soil; transporting, mixing, and applying multiple kinds of chemical fertilizers, insecticides, and herbicides; and tending long rows of a single commodity crop.

Laboring on Dona created a kind of rural anomie among Casiavera's laborers wherein their agricultural and ecological knowledge was replaced with a curtailed knowledge of repetitive agricultural work. An industrial labor discipline, involving precise measures of time and movements, was instilled in the workers with supervision.[27] There was no autonomy in when and how the work was done. An overseer, a man from the city who studied ginger cropping in Pakistan, mandated labor schedules. Work was atomized and compelled; spending all day on the plantation removed the laborers from a life of collaborative work.

Along with the plantation's logics of work and time came the way industrial agriculture attempts to turn the biophysical interactions between humans, plants, and machines into a biological machine with interchangeable units of germplasm, agrichemicals, and labor. It was an attempt to co-opt biophysical processes to accumulate wealth. If agroecology is diversity, indeterminacy, messiness, and collaborative work, monoculture is radical simplification, strict discipline, order, and atomization of both biological and social life.

At their most convoluted, industrial agriculture's large-scale monoculture plantations are said to free humans from the constraints of nature. But for all the technical innovations to disconnect capital from dependence on ecology, monoculture plantations are still tied to the biophysical constraints of growing plants. These remain relations between and among nature. Even as projects of selective breeding, hybridization, and genetic engineering have fundamentally remade the genetic material nature's biophysical and ecological processes act upon, more-than-human processes remain in control on the plantation.

After three or four years, things began to go wrong at Dona. According to Abril, this happened when Monica got a criminal type (*preman*) from the provincial capital to work as an overseer after the previous one, a man from Java with experience planting the ginger crop, left. "Everything became a problem then."[28] Monica told people in Casiavera that she had used up all her capital and had no more money to fund another ginger crop. Sometime in the mid-1990s the land again went fallow. Teuling rented it to Chinese Indonesian tobacco agents, who paid outsiders and workers from Casiavera to plant a few small patches of the crop. But it mostly went unused.

Casiavera's residents were not quite ready to defy the plantation's boundaries, afraid of attracting violent state repression. But the New Order's control of the archipelago was weakening. The militarized bureaucracy had stifled much freedom and hope, but in the second half of the 1990s agrarian dissent grew more open and impassioned across the Bukit Barisan. By then, Casiavera's workers had experienced the limits of the plantation system themselves. Many of them were resolved to take back control of the land from Dona. Eventually these workers were able to construct a path beyond the plantation, first by occupying the land and then by bringing forth a more vibrant collection of life on it. It is to this remarkable reclaiming of land and life that I now turn.

PART II

Reclaiming

Without borders, without soldiers. . . . We rise up like heaven. Let friends work together to grow food for the sake of the future. Well-being is more important than oppression. Human and nature are equivalent.

Tanpa batas tanpa tentara. . . . Kita bangun seperti surga. Mari kawan bekerja sama menanam pangan demi masa depan. Kesejahteraan lebih utama daripada ketertindasan. Manusia dan alam itu setara.

—From the movement song "Utopia," recorded in 2011 by the Indonesian artist collective Taring Padi (Rice's Fang)

CHAPTER 4

From Dissent to Occupation

Reclaiming requires giving up whatever calm exists in the current order to create something new. New knowledge, identities, politics, and action combine to create new relations of land, work, and autonomy. In these projects, those who desire to reclaim the land, mine, factory, or housing development can't rely on others to improve their plight; they must shoulder the great responsibility and challenge of reclaiming themselves.[1]

The long effort of reclaiming in Casiavera began in 1993, quietly. After years of secret discussions, about forty families who lived in Casiavera as part of the local Minangkabau Indigenous majority, nearly all of them one-time laborers on the plantation, or "coolies" as they called their old job, started clearing land on the edges of the plantation. Individuals, married couples, and cooperatives began working small patches of land, planting cinnamon trees, and trying their hand at raising chilis and eggplants. Another ten or twenty families joined them over the next two or three years. The work to clear the tough grasses that grew up on the damaged plantation land was time consuming and difficult.

This first wave of people to occupy the plantation maintain that the company was not working the land at the time, but their efforts did catch the attention of plantation staff. Zed, then the village head of Casiavera, confronted company overseers. "I said to Dona, we will profit from this land, not you! We went up, full of bravery." During these early days, whoever wanted to work the land went up and began to clear, till, and plant it. According to Zed, "We followed our custom, it was collective land to be used by the people from the community, but it could not be sold or pawned (*gadai*)."

Aren, whose parents named him for the volcano where he was born, was a local activist who took to visiting Casiavera to support the occupation. Of the occupation's early days Aren recalled, "We were still underground (*rahasia*), we had to be very careful." After much deliberation they occupied the land, despite the risks. Aren continued, "It was still so hot. The military would come and look at us, with their guns. The government people too, doing surveys and making maps, asking us who was living there, who was using the land." Aren is now the provincial head of one of Indonesia's largest smallholder organizations, the Indonesian Peasant Union. With years of reflection about his time in the Casiavera occupation, Aren came to recognize the singular importance of the early discussions and consensus building in the subsequent durability of the reclamation. Aren elaborated:

> We mostly just talked, but that was very important, to start to talk. These [discussion] groups helped defend us and helped us maintain, just because they were there. We would get together.
>
> Then the community council, the village head, the community leaders, they all decided that we would take the land. It was not underground anymore. They came out and empowered (*berdayakan*) the people.[2]

A second occupation wave began immediately after the New Order crumbled in 1998. Some two hundred families went up to the land to join the sixty or so that had already occupied it over the preceding four or five years. About half of these new families included members who worked as laborers on the plantation, a smaller percentage than the families who joined the first reclaiming efforts. Most of these reclaimers were members of families who held lands down below, in the village, and used the collective land to add to their holdings.

Before the reclamation, many in this second wave had left Casiavera to work as street vendors and construction workers in Sumatra's burgeoning cities. A few had migrated to Malaysia to work on the oil palm plantations there or as nannies. After hearing about the start of the occupation, this group returned to Casiavera to reclaim the land and begin their lives as smallholders.

All shared an interest in reclaiming the land above their homes and perceived a strategic opening when Suharto fell from power. Only a few of the reclaimers were landless, and even if they did not have any individual claims to their families' lands, they were all members of property-holding families in Casiavera. Indeed, recognized standing as a member of a Casiavera family lineage by birth or marriage was the de facto requirement for access to a land plot. They shared common experiences as Casiavera locals, Minangkabau peoples, and fellow laborers at the Dona ranch and plantation, giving them a common ground for their reclaiming. Their deep sense of place, of belonging to the landscape, added to their feeling of justice denied.

High up on the volcano, the occupation avoided the attention of the government until January 1996, when Dona Company applied for an extension of its concession (HGU) to the National Land Agency. As part of the application, a team of technicians from the district National Land Agency arrived at the volcano to check on the concession. The technicians' first stop was the village head's office. There, the team informed Zed that the Dona Company wanted to extend its land lease for a second twenty-five-year period. As part of the extension process, the technical team said, they would be carrying out a few routine measurements of the land. Up on the plantation, the Land Agency team recorded clear signs of reclaimers' occupation of the plantation land.[3]

The technical team's visit marks the opening of a more public and confrontational chapter in Casiavera's land struggle. The occupation changed from this point on because reclaimers knew that for the first time the government had recorded evidence of their occupation. With their visit, the Land Agency staff also unknowingly provided Casiavera's residents with a critical piece of information. The team told the village head that Dona had applied for a *second* twenty-five-year concession. No one in Casiavera had ever seen the company's first land concession permit, but it was reasonable to conclude that if this second permit would be for twenty-five years, that the original land concession was also valid for twenty-five years. And many remembered vividly the exact year, 1968, when Mahmud Teuling arrived to take their land. Thus, by 1996 Dona's original land lease had already expired.[4]

The Casiavera Council called a series of meetings to discuss what, if anything, they could do. They were disappointed with the loss of their land and the New Order regime for not upholding what many considered a founding principle of the Indonesian revolution: the protection of their freedom to be free smallholders, not plantation laborers. At their meetings the council reached a consensus. They recognized the expiration of Dona's concession as an important strategic opening toward the reclaiming of their agricultural land. If the concession extension could be stopped, Dona would not have any legal right to the land. And without the company on the land, Casiavera's agriculturalists' case with the state for legal access to the land would be improved.

POSSIBILITIES AT THE END OF THE NEW ORDER

The expiration of Dona's land concession came at a fortuitous time for Casiavera's reclaimers. As memories of the late 1960s genocide that had brought Suharto to power dimmed and the New Order's command-and-control regime aged, dissent became more fervent and daring in the 1980s and 1990s. In the countryside, organized and coordinated acts of resistance to dispossession and enclosure became visible. By 1990 a number of problems stemming from development projects were being covered in the newspapers, not least the violent evictions of smallholders to make way for plantations, but there were also mentions of increasing trash, sewage, air pollution, deforestation, diversion of water from rice paddies, and the negative effects of mega-dams.[5]

One of the most vivid, contemporary accounts of New Order rural struggles is a 1991 calendar colored in ink, called the Land for the People Calendar. Two university students in Salatiga wrote and illustrated the calendar, an effort that landed the pair in jail. The twelve cases featured in the calendar sketch the wave of reclaiming that was unfolding. In Cidebung the children of peasants dispossessed in 1973 reclaimed seven hundred hectares of plantation land. In Situbondo thousands of farmers destroyed a coffee and chocolate plantation, reclaiming it to plant corn and soy. In Jenggawah and Jember migrant workers joined up with locals to occupy state-operated tobacco plantations. And in North Sumatra two thousand farmers demanded the return of one hundred thousand hectares of timber plantation land seized by a parastatal logging corporation at the start of the New Order.[6]

Since its founding in 1994, the Consortium for Agrarian Reform (Konsorsium Pembaruan Agraria, KPA) has maintained a land conflict database covering Indonesian newspaper and activist reports. From 1970 to 1999, the organization recorded 1,753 cases of land conflict in which nearly one million households were evicted or dispossessed of their lands, encompassing a total area of some ten million hectares.[7] Land conflicts were nearly evenly distributed between the rural and the urban, with contentious large-scale plantation expansion (20%), mining and dam building (9%), logging operations (8%), and forest conservation and protection (3%) in the archipelago's fields and forests. In the cities, conflicts centered mostly around infrastructure, factory, and housing projects (34%). The frequency of agriculturalist counter-dispossessions increased from 1980 to 1999, including reports of unarmed workers occupying land slated for industrial development, blockading plantations and mining roads, committing acts of arson against company assets, and reclaiming land.

State policy was to attack, discredit, and criminalize the growing land occupations. In 1991 Suharto told farmers to not become "resisters" (*kelompok mbalelo*). Land occupations were especially subject to sustained denunciation. The media repeated the government's labeling of these actions as "land looting" (*penjarahan tanah*). The land occupations were called "illegitimate" acts of "civil violence and anarchism."[8] The farmers who participated in mobilizations were described as "unduly argumentative" (*waton sulaya*), "jumping bedbugs" (*kutu loncat*) because the same movement leaders showed up at many of the disputes, "sate sellers" fanning the flames of land conflict just as a merchants fan the flames of their sate grills, and prostitutes (*wanita tuna susila*). The minister of home affairs took an even more provocative step, labeling reclaiming movement members communists.[9] Yet land occupations and reclaiming increased, as did the solidarity protests of workers, students, artists, and journalists in the streets.

LETTER WRITING FOR FREEDOM

People in Casiavera were by no means isolated from the agriculturalist mobilizations that began to sweep across the nation at the end of the New Order. Casiavera's residents acted with the knowledge that they were not the first community to try to reclaim their land from a New Order–era plantation head-on. Their experience was reflected back to them in the grievances motivating other protests. Still, there were many violent repressions of agriculturalist mobilizations, and very, very few places where reclaiming brought agriculturalists well-being.

Given the risks, the Casiavera Council chose to proceed with caution. The council began not with direct action or protest but by writing a series of letters to the National Land Agency, for the simple reason that the agency would decide on the extension of the plantation concession. Theoretically, the Land Agency held the power to return the legal right to the land to the community, although no one had heard of that happening. The ensuing letters from the council are rather remarkable for the way they exposed the Dona Company's harmful effects and at the same time interrogated the very underpinnings of the state's authoritarian, developmentalist logic.

In the council's first letter in February 1996 their demand was clear: they wanted the Land Agency to reject the plantation company's application for an extension of the land concession and place control of land with the community government, to be used by the people who lived there. But unwilling to come right out and condemn their state-sanctioned dispossession, the council was more comfortable starting with a bureaucratic appeal for transparency. The title of their first letter underscores their delicate approach: A Respectful Request for a Photocopy of the Land Concession Permit and Map of the Collective Land That Is Currently Used by the Company.[10]

The delicate tone of the council's first public written appeal to the state belied Casiavera's initial position of weakness. The council's bureaucratic request—the sharing of a government permit and map—was a tenable entry point to the conversation about who should control the land. Taking the form of a state-sanctioned objection, the council's letter contained an element of submission to authority, but it also opened with a rather bold statement:

> We as the legitimate owners that have descended from this collective land do not clearly understand the status of the land that is part of the land concession that the company holds and there are not any defined borders of this land concession. This unclear status of the land and the undetermined limits of the company land creates an uncertainty that could lead to conflicts that we do not want even though the company has taken control of the collective land in its entirety and puts pressure (*tekanan-tekanan*) and intimidates the people of the community that want to use lands that are outside of the HGU [land concession].[11]

With their opening remarks in what would become a long line of communications to the government on the status of the Dona concession, the council stated that they were the "legitimate owners" of what they called the "collective land." And they took up a point that they repeated for years in their letters: the plantation's land concession boundaries were ambiguous. This was the council's use of discourses of mapmaking and spatial information as a strategic point of entry to their land problem with the state. It is likely that the council was aware that no plantation boundary map existed. The original 1968 land lease did not include one. But it was increasingly expected in the 1990s that state land claims relied on the legitimacy of maps and mapmaking. The council raised the absence of this spatial information as a strategic opening to challenging the state and Dona's position. For them, the council wrote, "the situation was unclear." A general "uncertainty" existed. The concession boundaries were "undetermined."

The sharing of permits and maps may have been the council's stated intention in writing to the National Land Agency, but even in this first letter the council went beyond this initial administrative request. They detailed the company's intimidation and persecution of agriculturalists as wrong-doing. Then they applied the New Order's own dogma of development against the plantation company. The council wrote that the company was not making productive use of the land. According to the council, the company "neglected" (*diterlantarkan*) and "underutilized" (*kurang dimanfaatkan*) the land. By not bringing development or modernization, the company was reneging on its commitment to the Land Agency to make use of the land for the benefit of the national economy, the council claimed. In this way the council introduced claims crafted to undermine the justification of rural development itself. The plantation company had legitimized its control of the land with ideas of development and modernization, namely the efficient and productive use of the land. But for Casiavera's dissident scribes, the plantation was none of those things; it was inefficient and wasteful.

Here, the council was attempting to use one of the most potent ideas of Suharto's New Order for their own purposes. In 1994 Suharto had issued the *President Instructions on Left Behind Villages*. This was a set of policies that the national government had developed closely with the World Bank to identify settlements with a higher than typical number of families living in poverty and award these villages special development grants. The aim was to spur economic growth specifically in these locations. That same year, a World Bank technical team declared Casiavera to be one such "left behind village" based on its low average income and limited agricultural resources. The council seized on the rationale of this program, arguing that if "President Suharto himself" identified their community as "left behind" and "poor," and Suharto had issued a policy to do something to change this unfortu-nate fact, then Casiavera needed a different form of development. The coun-cil's letters to the government argued for a new trajectory of development that would involve returning control of the land to Casiavera's residents for smallholder agriculture.

This first letter included one final, crucial detail. At the end of the letter the council wrote that Casiavera's leadership had already taken the decision to allow smallholders to cultivate lands that the council believed to be "collective land outside of the plantation concession." This was Casiavera's first public act of reclaiming. The council did not yet challenge the authority of the company and state to control the core of the plantation, but they did inform the Land Agency that people already occupied areas at the edges of it, lands that the council stated to be outside of the state's land lease.

The council also wrote that Casiavera's reclaimers had taken action so that "members of the community that are below the poverty line" could cultivate what they called "high economic value crops." These "high economic value" crops would replace grasslands on the abandoned plantation. According to this argument, the smallholder's efforts would be more productive, and in the long run, more beneficial for the nation, than the company's form of management that left the land "underutilized." The agriculturalists had acted not in rebellion, the council argued, but only to realize Suharto's own vision of economic improvement and poverty alleviation for "left behind villages" like theirs.

Living in a milieu of pervasive New Order violence, it would not have been unreasonable for the council to expect a reactionary response to their letter. Instead, they got a simple memo from the local Land Agency office. The communication did not acknowledge the council's statements of grievance or the ongoing occupation. The Land Agency wrote only that if the council wished to access a copy of Dona's concession permit, they would have to make a request to the provincial Land Agency office. In response, the council expanded the recipients of their next letter to include nine state officials, including the head of the provincial Land Agency, along with the district head, regional head, and Casiavera's representative to the provincial legislature.

In this second letter the council again led with their administrative request for photocopies of the land concession letter and clarification of the plantation's boundaries. And again, they quickly moved to the more serious issues of the occupation, saying that they were only motivated to write "because today Casiavera residents beg us to be able to cultivate the collective land for poverty reduction (*pengentasan kemiskinan*), as called for by the Government today."[12]

After nearly a year the council, still waiting for a response, escalated their letter-writing campaign. They wrote a third letter, this time to the governor of the province, with copies to more than thirteen state officials, with this more assertive title: Pleading to Request Permission to Cancel the Dona Company HGU Extension on the Collective Land.[13] They began by expressing regret that their previous letters had been "ignored," "disregarded," and "sidelined," while the National Land Agency was much more responsive to Dona's request for an extension of the land concession. "Without any other options," the council laid out for the governor a new set of arguments against the company. The council wrote that in their analysis the land was never state land and so the original 1968 concession was not valid. According to them, the land had been the community's collective land (*tanah ulayat nagari*) since "time immemorial" (*sejak dahulu kala*), managed by elders (*ninik mamak*) as a collective inheritance (*warisan*).[14] It was upon this basis of historic possession and use that the council requested the governor to work with the Land Agency to reject the land concession extension. The council sought approval to use the land as smallholder plots (*ladang*) in the plantation's stead, because they "very much need[ed] the land . . . to use as a pilot project to support the Government's program to alleviate poverty."

The council's inversion of the New Order's ideology of development and modernization was pragmatic; couching their demands in the language of poverty alleviation and bureaucratic procedures was almost quotidian. But writing letters to gain control of their land was anything but mundane. Indeed, their effort broke from a decades-long practice of peasants seeking to avoid the attentions of an authoritarian state and by extension staying out of government records and archives altogether. Pen in hand, the council was carefully going about writing their arguments and history of the land into state records, on their terms, and using the language they believed was most likely to draw government support away from agribusiness and toward their own, dissenting perspective.

While the council wrote letters to the government, they also reached out to Casiavera's large migrant diaspora for movement support. The council retained three lawyers in Jakarta and one in Padang, West Sumatra. One of the lawyers was the chair of the Casiavera Family Association, which held regular meetings in Jakarta, Bogor, and other Javanese cities where migrants from the community had settled. A power of attorney agreement dated March 3, 1997, requested the lawyers help document the case against Dona's concession extension and solicit support from the Indonesian Legal Aid Foundation, Commission on Human Rights (Komisi Hak Asasi Manusia), and legislators in the National House of Representatives (Dewan Perwakilan Rakyat).[15]

Taken together, these advocates constituted a powerful diasporic influence that drew on legal and cultural expertise to advance the reclaimers' cause. Casiavera's smallholders were able to build this coalition through their strong ties to migrants from their community, as well as their long history of supporting and nurturing intellectual achievement. These were resources not at all typical of other land occupation movements, and they played a significant role in the council's successful elaboration of a critique of the Dona Company's justifications for an extension.

Shortly thereafter, the council received a copy of a telegram from the West Sumatra division of the Body for the Coordination of National Stability (Badan Koordinasi Stabilitas Nasional Daerah, Bakorstana).[16] With this telegram, for the first time the council had reason to think that their efforts to end Dona's control might be successful.

Bakorstana was a much-feared military intelligence agency. Its officers were men and women from the military, police, and state attorney's office. Agency staff primarily concerned themselves with arresting government critics, students, and anyone else believed to have links to the banned Indonesian Communist Party. Suharto had established the agency to replace the notorious Restoration of Security and Public Order Command (Kopkamtib) that directed the early New Order pogrom and became the institutional basis for military rule of the nation.[17]

Incredibly, the Bakorstana telegram concluded that the council's claims were accurate and reasonable. Deviating from its established role as a repressive force of state violence, in this case Bakorstana backed up the council's assertion that the company's land concession wasn't valid. The letter also mentioned that the agency planned to carry out additional "checking" and "research" and would work to settle the issue according to applicable procedures. It was the first state agency to acknowledge any validity to Casiavera's case. Bakorstana's support can only be understood with the knowledge that Casiavera's reclaimers worked especially hard to sway one Casiavera migrant to support their movement: the very officer in the Bakorstana office who authored this letter.

Even from within the ideological concrete flume that was the authoritarian New Order, the Bakorstana letter attested that there remained space—however constrained—for ideas of community and moral behavior with those in one's social orbit. Because the Bakorstana officer was from Casiavera, grew up with those who suffered at the plantation, and came to believe that they had been wronged, he was in a position to use the power of his office to aid Casiavera's reclaiming. Operating in a way that was contradictory to state policy, this officer of the law and armed forces sought to bolster Casiavera's reclaiming movement.

The undermining of the state's position and power to control Casiavera's residents was a kind of rightful resistance, where within a structure of domination there was room to accommodate a popular resistance movement.[18] By locating a sympathetic elite officer and persuading him to support their claims to the land, Casiavera's reclaimers cultivated resistance within an agency of state repression. Even while the reclaimers decried their state-led dispossession as illegitimate, they made use of a collaborator with access to state power to accomplish their reclaiming.

TO THE BLOCKADE

Beyond sympathizing with his fellow community members, there is another reason why the Bakorstana officer may have written his letter of support: direct action. Just two months before the release of the letter, in late 1996, Casiavera's reclaimers started a protest blockade. The action followed a meeting the regional head had called with Dona staff and Casiavera residents in order to "avoid undesirable events." By the Land Agency's own estimates at the time, roughly one hundred families were cultivating the erstwhile plantation land. To prepare for the meeting, the Land Agency instructed Dona to measure the boundaries of its plantation and come to the meeting ready to discuss these boundaries. The day before the meeting, a Land Agency technical team visited the plantation to prepare measurements of their own.[19] The Land Agency team, unaware, walked right into a full-blown agriculturalist blockade.

Hundreds of people confronted the Land Agency team, standing and sitting in the road that ran up to the plantation. Spokespeople in the crowd that day told the team they would not be allowed to do their work. The protestors' numbers and piles of wood and burning tires at the blockade signaled their intention to prevent any government survey of the land. They also suggested that someone had leaked the plan for the Land Agency's survey to the protestors.

Tia, mother of two and one of the largest landholders in Casiavera, was there that day. She told me, "The news spread: The government was coming back. We blocked the road; people carried their air-rifles and machetes. We no longer trembled with fear. No, we were not afraid anymore. The people were reckless (*nekat*), we were not even afraid of death."[20] Others used the same word, *nekat*, to talk about the time they began reclaiming the plantation too. They acted regardless of the risk. In their telling, they channeled a long pent-up rage and held close their belief that although their actions were illegal, they were just. More than a bit of recklessness was needed to join the blockade, a form of protest that has long been fiercely criminalized and repressed in Indonesia. Moving up to the land and declaring it theirs exposed Tia and hundreds more from Aren to being labeled criminals. The risk of injury or arrest was real. Beyond suffering outside in the rain, heat, and cold, there was the not impossible eventuality that she would have to stand in front of an oncoming group of armed soldiers.

Protestors told me there was not much exuberant singing at the blockades. Most parents left their children at home, if they were fortunate enough to have a home, just in case militarized police started a violent riot or hired thugs showed up with machetes, sharpened stakes, jerry cans of diesel fuel, and lighters—as they did when they burned down activists' homes at another Sumatran Indonesian Peasant Union land occupation that I visited. While Tia and others came to think of their blockade of the plantation as reckless, they did not dwell on the fear of repression. Standing in front of the bulldozers and police involved despair, long pent-up anger, and a bit of fatalism, but joy and exhilaration as well.

As a Minangkabau woman enmeshed in a complex web of family ties, connections to place, and economic activity that ran scores of generations deep, Tia joined many more women in Casiavera in protest. Their leadership and perspective were fundamental to the reclaiming movement. It was these women who helped to maintain family-lineage affairs across generations of dispossession. And it was these women who held the power to control community access to land across Casiavera for all of their relatives, or what they called their "sister's children" (*kamanakan*).[21]

Zed, the village head, recounted his role in organizing the blockade: "By the end of 1995 the plantation concession expired. I went first to the Bupati, then to the Land Agency in Padang. My neighbor, he helped me get access to some of the government documents. But we did not have any path forward. So, we went, as a mass, all the community, up to take the land. We went to kick them [the company] out."[22] In framing the action as the result of not "having any way forward," Zed was alluding to the fact that the protestors did not see any access to legal channels to get their land back, so they took to the blockades.

In late 1996, when Zed heard the Land Agency was planning to come, he gathered the leaders of the Casiavera Council and their youth organization for a visit with the regional head, who advised them that if they did not agree with the concession, "Do not let the company do their work then." It was a signal that sympathies for their cause had spread up into higher ranks of the state. On December 12 Zed joined the protest on the road: "We made them [the Land Agency team] disappear!" With such potentially insurrectionary direct action, suddenly anything became possible. When Zed, Tia, and the others raised a barricade, they walked out through the door of state custodianship and into unknown, more autonomous forms of life.

Casiavera's blockade was an escalation. The workers turned activists were no longer engaged in a quiet land squat far from the centers of state power. While their occupation was mostly an effort to improve their position without provoking state reaction, their blockade was a bold counteraction against the state's claim to control the land. The action was an inflection point from obedience to dissent, very close to open rebellion. This was a much riskier form of direct action than respectful letter writing, for Sumatran plantation and logging company blockades are often met with arrests and violence from riot police; company-employed security; and pro-plantation militia, thugs, and enforcers.[23]

At ten o'clock in the morning that next day, Zed joined the council to meet with the Land Agency and other state officials. According to the regional head's meeting invitation letter, just 7 people were invited. More than 150 showed up. Attendees from Casiavera told the agency representatives that they did not agree with measuring the land or Dona Company's control of it. Many left the room midway through the meeting in disgust with the Land Agency staff's responses. Those who remained eventually struck an agreement with the agency. They would allow agency staff to measure the land and guarantee the safety of the technical team while they did their work if the survey made note of the number and location of reclaimers that were cultivating the land.

The Land Agency staff agreed to these conditions and made two further commitments to the reclaimers that brought an end to the blockade: First, it was against state regulations to renew the concession if there were outstanding disagreements between the company and local residents. Second, if the company and community members could not reach a settlement to their conflict within one month of the meeting, the concession could be canceled according to Land Agency procedures.[24]

Backed up by Casiavera's blockade on the volcano, the council began to increase the strength of their rhetoric in their appeals to the Land Agency. In their next letter, dated April 17, 1997, the council listed thirty-four points explaining why the land should be returned to them, starting with the fact that Dona had "seized" (*diramapas*), "occupied'" (*diduduki*), and "illegally owned" (*dimiliki secara ilegal*) the land since 1968.[25] The council claimed Dona had gained access to the land by "manipulating data and supporting documents" and was never actually economically productive. The Dona Company had played "tricks" by buying ginger from other places and taking it by truck to their plantation to make it look like production was underway in order to convince the Land Agency that Dona was in fact, operating. But really, the council claimed, the company was bankrupt.

The council went as far as to say, "Dona did not use the land." As such, the council demanded the end of Dona's "occupation/confiscation" (*pendudukan/perampasan*). Again, the council took up the language of the New Order, and again the council reversed its operating logics. Underpinning the movement was people's belief, counter to that of the New Order, that it is smallholders, not companies, that are the most economically productive land managers. And that companies, not smallholders, are the real squatters and occupiers.

Their legal expertise now deepening, the council for the first time referred to specific Indonesian law to support their reclaiming. They argued that the Dona land concession was in violation of the single most important agrarian policy, the 1960 Basic Agrarian Law. According to the council's legal analysis, all concession lands must be managed in accordance with the investments and engineering typical of modern development. But the council claimed that Dona's concession was neglected (*tidak terurus*) and only a "weedy field" (*padang ilalang*): "Since 1970 until now the illegally occupied land has never been as productive as it should, which is contrary to the principles adopted by the Basic Agrarian Law 5/1960 (that the land should not be neglected)."[26]

They went on to argue that the original incorporation of Dona Company as a Dutch colonial-era limited partnership corporation (*Commanditaire Vennootschap*) was "not a legal entity in Indonesia recognized by the Basic Agrarian Law and so they should not be recognized as such." And the council used a little-known government regulation, number 10 of 1961, to argue that the company did not fulfill its requirement to publicly announce the creation or extension of the concession. For the first time, the council also made the case that their collective land was never part of the colonial-era W. H. Samuel *erfpacht* land lease, number 203 of 1929, that Dona's HGU was based on. They claimed that they could make this argument because they had found *erfpacht* documents supporting their dissenting argument at the National Archive in Jakarta and through a hired representative who visited government archives in Holland.

Along with this greater legal prowess came an increased concern about the specificity of the documents that supported Dona's concession. The council noted that they had never seen any map of the concession, questioning whether one existed. They argued that the public document that underpinned the concession, the land use permit, did not give any meaningful instruction on the location of the land-lease boundaries. The cardinal directions and the landmarks the permit referred to were said to be mixed up and contradictory.[27]

In the freewheeling remaking of agrarian space in the early New Order, plantation owner Mahmud Teuling had used his personal power to establish control of the landscape. Without a map or location clearly specified, Mahmud Teuling was able to decide what land he wanted and take it. In this dispossession regime, elites' ability to grab land was dynamic and flexible, as were the plantation boundaries. But thirty years later, the council used this original geographic opacity in the land use permit to their advantage.

The council also deepened their moral justification. They said they were carrying the "mandate" (*amanah*) to "relentlessly struggle in search of justice" (*tanpa hentinya berjuang terus mencari keadilan*). The council claimed the ethical high ground and access to fact and truth (*benaran*):

> In good faith the leaders of the Council have tried to resolve this problem in accordance to the law. It is very naive that our people have been harmed in the interests of a single person. Finally, after our people have sought their rights and justice if one day they rebel (*berontak*) and riot (*terjadi kerusuhan*) who will be responsible?[28]

Their argument was draped in the moralistic language of social mobilization. It took on the emotions of protest. Casiavera's reclaimers found faith in their protest action because they claimed access to a higher, more virtuous understanding of reality. They sought to throw off everything about their dispossession that enclosed and constrained them. And if they were prevented from obtaining this justice and freedom, the council said, it would not be their responsibility if the reclaimers turned to rebellion or riot.

The council constructed the reclaimers as not only righteous but also a potentially destructive force if the concession extension were issued. The council concluded with what by then was a familiar demand, that the land be returned to the people: "For twenty-five years the community of Casiavera has been oppressed (*tertindas*) and their land has been seized to benefit private interests. This land should be used by people for gardening, livestock and agriculture."

Most of 1998 passed without a reaction from the government. Finally, on the first of October, the provincial Land Agency wrote to its superiors in the Ministry of Agrarian Affairs, in Jakarta, asking for direction. In a letter titled The Problem with HGU No. 1/The Elders of Casiavera Demand Its Cancellation, the provincial land agency revealed a startling turn of events; at some point in 1997—unbeknown to anyone in Casiavera—the provincial Land Agency had extended Dona's concession for twenty-five more years.[29]

After issuing another land concession lease, the provincial land agency officials sought guidance from their state superiors on what for the agency had become a "problem." It was a problem because a few months earlier, in August, the community had escalated their tactics, staging a demonstration outside the West Sumatra governor's office, where they demanded the cancellation of the Dona Company leasehold. Casiavera's reclaimers brought a rare open protest to the streets, forcing the governor's attention for at least a few hours.

In its letter, the Land Agency reported that staff had completed their survey of the land, where they found the community controlled 90 percent of the concession. The technical team also noted that the land was under the management of the council, who vowed to not give it up. There was reference to a Dona "base camp" and a number of company staff on the land, but even with these resources the company was unable to block agriculturalists' access. Even though the provincial Land Agency concluded that the land was under the day-to-day control of Casiavera's agriculturalists, in its letter the provincial Land Agency office declared that it was the plantation company that should control the land. The agency concluded the letter with a suggested solution to the problem: to register the identities of any reclaimers who continued to cultivate the land and instruct them to vacate their plots. The process would be managed by the National Land Agency's Land Investigation Committee, with the power to suggest that the district police arrest the reclaimers.[30]

RECLAIMING THE LAND (AGAIN) IN 1998

Despite their considerable political effort, Casiavera's agriculturalists were not able to prevent the extension of Dona's state land lease for twenty-five more years. But even after Dona restarted efforts to control the land in 1997 with its extended land lease from the Ministry of Agrarian Affairs and an infusion of investment capital from Indonesian agribusiness, the reclaiming held.

The critical day came when company staff rented a fleet of heavy equipment with the intention to bulldoze agriculturalists' plots. Dona's director solicited support from the nearby Indonesian Army base to protect the bulldozers while doing their work. The day the bulldozers and tractors arrived in 1997, soldiers escorted the equipment. A crowd met the machines at the top of the road that led to the plantation. The reclaimers had walked out of their homes at the top of the volcano and simply sat down, blocking the company's only access point into Dona. The uneasy standoff cooled when it was clear that the protestors had called the army's bluff. The soldiers were not willing to take their weapons off their shoulders and use them to disperse the blockade. Zed explained, "By that time the dual function of the military, ruling not just war but also politics, had ended. Thus, the Army would not fire on us. They let us demonstrate. The Army retreated." Zed saw this second, definitive blockade through to the end, sweating for days under the equatorial sun.[31]

Women and men, middle-class smallholders, and impoverished migrant landless laborers joined together on the blockade in an affinity politics. They worked together because of a shared desire for shutting down the machines of plantation dispossession and destruction, not passionate support of any leader, party, or union—a kind of diverse and intersectional mobilization of the multitudes.[32] Relations were built up through shared desires to overcome exploitation and the willingness to become accomplices.

Finally, after five years of street protests, road blockades, occupying and cultivating land, and a letter-writing campaign to solidify their standing with the state, Casiavera's activists left Dona Company without any practical way forward; ever since this second blockade, the company has not had a physical presence on the land.

Land occupations are a foundational peasant strategy for survival among dispossession and landlords in the countryside. The durability of this peasant tactic attests to its emancipatory potential.[33] Although superficially similar to a classical peasant occupation, the reclaiming of the moribund plantation bore little resemblance to the revolutionary, class-based politics of the peasant movements of the 1960s and 1970s. Unlike these peasant wars, there was a marked non-violence to the agriculturalists' struggle in Casiavera. The refusal to engage in violence was owed, in part, to the contingencies of the Bukit Barisan, where there was no private, land-ruling class accustomed to enforcing its control over those less powerful with pugnaciousness and violence, as was so often the case in systems of land control with feudal genealogies. The lack of armed insurrection was also due to the fact that with their resistance, the reclaimers sought to challenge their rulers but not overturn them.

The reclaiming movement in Casiavera was not a radical Marxist class politics or anti-state position. It was a pragmatic workers' mobilization. Even while reclaiming unfolded at the edge of a bloodless revolution that would bring about the end of one of the twentieth century's longest dictatorships, the land struggle in Casiavera stopped short of the complete repudiation of the New Order regime. Casiavera's activists and leaders did not demand the total restructuring of power, only that the exiting regime treat them justly. Their concerns were more personal, less national. Their resulting political action was not outward looking, nor was it aimed at reform of national ruling institutions. As such, reclaiming action was narrowly focused on microscale, place-based politics aimed at gaining control of the land. The occupation was the end goal itself. Smallholders' greatest hope was that the company and the state would simply leave them alone to cultivate the land on their own.

All told, hundreds more smallholder mobilizations in Indonesia joined the one in Casiavera to constitute a collection of reclaiming movements. Since the fall of Suharto's New Order, the Indonesian Peasant Union alone has supported some twenty land reclaiming movements in Sumatra.[34] Unfortunately, state and corporate security forces dismantled many of these outright with criminalization and persecution. Others have collapsed under the joint weight of external repression and their own internal contradictions. Too many reclaiming sites remain dominated by ideologies and structures of capitalist control and work, becoming sites of unequal land control and commodity cropping.

Often, smallholders erroneously seek to replicate agro-industrial monocultures of oil palm and rubber on land that can be freely bought and sold as private property, that is, smallholder plantations. Individual property titles and smallholder plantings of rubber and oil palm on reclaimed lands are especially capable of generating a highly unequal local economy. Through the influence of government agencies, local politicians, and organization leaders, the momentum of the already existing, very unequal, and repressive social order can carry through into reclaiming movements. The total effect is that the ecology and distribution of many movement landholdings form into ecosocial structures that smallholders were originally seeking to overcome with their reclaiming.

An example is what happened at Garut, in West Java, where thousands of agriculturalists occupied a rubber and teak plantation in 1997. A decade after creating individual property titles to fifteen hundred hectares of land in the onetime plantation, 10 percent of reclaimer households were impoverished and landless, and less than 5 percent of landowners controlled nearly 40 percent of the reclaimed land. These rich smallholders planted smallscale rubber monocultures, effectively smallholder plantations built along the agribusiness model, accumulating plots from other reclaimers.

A similarly troubling, unequal reclamation unfolded in Lampung, South Sumatra. There, at the end of the New Order, about one hundred smallholder families occupied a logging concession that the state-owned logging company Inhutani operated. A decade after gaining legal recognition of the land from the National Land Agency in 2003, nearly one-third of families that originally joined the reclaiming movement were landless and a single businessman controlled nearly one-third of the land as an oil palm mono-culture on land he had bought up from reclaimers.[35] At both the Garut and Lampung reclaiming sites, individual property titles and plantings of rubber and oil palm commodity crops brought forth a highly unequal local economy, allowing a few landholders to use their wealth and power to accumulate plots from their neighbors.[36]

The Pawartaku reclaiming movement in East Java has struggled with a different kind of trouble, one that is more political than economic. In 2004 a smallholder cooperative with its origins in Indonesia's post-independence plantation workers' union Sarbupri reclaimed a tea and coffee plantation. In 2012 the cooperative accomplished legalization of land title for 320 households with the National Land Agency. Organization members said this process was "just," because it ensured that the families who sacrificed the most during their long struggle for the land were largely able to gain land titles. Plots of 0.6 to 1.75 hectares were formalized in a participatory process of determining who deserved land (in exceptions, an organization leader gained 5 hectares and four customary leaders gained 2 hectares each).

Yet after legalization, the reclaiming organization faltered. One key organization leader capitalized on his influence in the community to move into the formal political sphere, joining the most authoritarian of the major Indonesian political parties, Gerindra. His connections with the Indonesian Peasant Union notwithstanding, in 2014 the leader ran for a regional legislative seat as a Gerindra member, a party with a pro-agribusiness and neo-fascist agenda that directly works against smallholder economies and agrarian justice writ large, saying he joined because of political pragmatism and the fact that Gerindra was the only party that did not require him to contribute funding to the party.[37]

By surrendering to the state's development agenda and neoliberal capitalist logic, this leader of this reclaiming movement threatened to become part of the political forces working against reclaiming movements. The problem is often repeated; political, military, and agribusiness efforts of depoliticization and assimilation alter reclaiming movements, allowing localized authoritarian relations to manifest. When the powers that be co-opt reclaiming movements, the forms of leadership and politics that emerge can work against the ability of movements to achieve any systemic expression.

The harshest, most brutal reclaiming failures have occurred at sites where reclaimers refuse to enter into state processes of settlement legalization at all, instead seeking more collectivist smallholder communities that exist outside of state rule. For their attempts at reclaiming a role for cooperatives, egalitarianism, and self-sufficiency, the police, military, and militias often attack these experiments in reclaiming. The Fajar Nusantara Movement of Javanese migrant back to the landers in Kalimantan is unique for the way that they constructed intentional migrant smallholder communities underpinned by an ideology of food sovereignty, autonomy, and a unitarian, universalist religion. It is also distinguished for its suffering at the hands of reactionary agrarian organizations and the state. Over a two-month period in early 2016, police and the military forcibly evacuated roughly eight thousand Fajar Nusantara members from their smallholder settler communities after gangs—identified only by their yellow headbands—attacked them. The victimized movement members were then sent to one of at least six unofficial detention centers in Jakarta; Yogyakarta; and West, Central, and East Java, from where they were turned over to local officials or returned to the custody of relatives in their hometowns.[38]

The authoritarian social formation made up of the police, military, mass media, and other organizations saw these reclaimers' commitment to communal living outside the control of the state as a destabilizing, threatening social influence, and so they directed their significant resources at dissolving the Fajar Nusantara communities. Without historical ties to the land or special Indigenous protections afforded under Indonesian law, the peasant union occupation was defined as illegal without exception by the state, which brought down its full force, culminating in multiple rounds of arrests and evictions of smallholders and movement leaders. In addition to state forces were the thugs and militia of the agrarian underworld, acting to add a layer of intimidation and fear on top of the state action, exacerbating the terror and danger settlers confronted.

In Casiavera, agriculturalists' reclaiming largely struck a more sympathetic note with the authorities. Reclaimers there have been spared the harshest persecutions. Their peaceful, reasoned justifications for reclaiming resonated with observers. Indeed, Casiavera's reclaimers found supporters across the government. In them, West Sumatra's authorities saw themselves. Although the state's economic development positions subsidize agro-industry and are antagonistic to smallholders, outwardly officials make great efforts to be seen as supportive of smallholders. Many in government are fond of saying, "We work for family farmers," even as the policies they bring into being too often work against smallholder agriculturalists. In many ways, Casiavera's reclaimers represented the peasant ideal for the powerful, the kind of rooted-in-place smallholders who independence and early republic leaders held up as the foundation of a post-colonial Indonesian society.[39]

The reclaimers also benefited from being part of the local Minangkabau cultural majority. The mostly Javanese bureaucrats working in the National Land Agency office in Jakarta did not see Casiavera's reclaimers as one of the minority Sumatran Indigenous nations like the Jambi people, who have long been marginalized and constructed as the other in mainstream Indonesian society. In contrast, Minangkabau peoples have held influential roles in the politics, theology, and literature of the Indonesian nation from its birth, despite making up not more than 1 or 2 percent of the population of the nation. Many of the leading statesmen and intellectuals of the early independence era, especially, identified as Minangkabau, including the first vice president of the republic, Mohammad Hatta, and the first prime minister, Sutan Sjahrir, among others.[40]

Without doubt, national bureaucrats awarded Casiavera's reclaimers a measure of sympathy and respect because they were Minangkabau, even while the state systematically marginalized many hundreds of other Indigenous peoples across the archipelago, denying these peoples the right to cultural expression and territory. For the Minangkabau, one of the benefits of their central role in the Indonesian revolution and early republic has been that mainstream Indonesian conceptualizations of them include a right to control their own lives and territories across a region of upland Sumatra that remains known simply as the Minangkabau. This acceptance of Minangkabau sovereignty was partial—state and corporate dispossessions unfolded here too of course—but it was nevertheless central to Casiavera's reclaimers' effort to take their land back.

For local Minangkabau elites and leaders, Casiavera's Minangkabau reclaimers did not threaten their control and power over society the way that the migrant Fajar communities did, which triggered local elites' long-ingrained fears of an influx of Javanese migrants displacing and dispossessing locals. And even while Casiavera's reclaimers elaborated an anti-plantation politics and egalitarian, collective land control, they did so in a way that was generally open to government collaboration and approval.

In 2012 Gunawan Wiradi, then eighty-three years old, visited Casiavera to receive a delegation of Via Campesina representatives. One of the few critical agrarian scholars to survive working through the New Order, Wiradi wrote two decades ago about the kind of direct-action agriculture Casiavera's residents carried out as "land-reform by leverage."[41] After a quiet walk through Casiavera's agroforests, Wiradi sat in the yard of one of the reclaimer's homes to explain the importance of what he had seen. For Wiradi, the way reclaimers used the land was a moral appeal to the state. Functionaries were forced to either endorse the occupation or repress it. And while the state wavered for years, reclaimers began to realize immediate gains. As Indra Lubis, a longtime friend of Wiradi and regional organizer for the Indonesian Peasant Union, has written, land occupations are a central movement focus because "with land occupations we achieve small victories and that's how we grow. We must keep hope alive with actions and small victories at the grassroots level . . . not just create or dream of a utopia to be won at the international level."[42]

Lubis's occupation work is a continuation of the now two-decades-long Indonesian reclaiming movements in ideology and tactics, as outlined by some of the movements' earliest organizers and thinkers in their *Reclaiming and Peoples' Sovereignty*, in which they called for move reclaiming action (*aksi reklaim*) precisely because occupations are not merely symbolic (*gagah-gagahan*) acts of protest.[43] While reclaiming is indeed resistance to injustice, it is also pragmatic because it is predicated on gaining access to land immediately, with occupation if needed, to provide smallholders and the landless an opportunity for livelihood as soon as possible.

Occupation, land reform by leverage, squatting: reclaiming as rural workers' act carries many names. In Casiavera it became agriculturalists' most effective political expression. Casiavera's activists brought together land occupation, blockades, and a letter-writing campaign to create a repertoire of confrontation that was both a potentially destructive force and a less radical effort to establish protestors' rights to citizenship under existing state rule. Their movement thereby collapsed forms of militancy often held to be disparate as either potentially destructive and revolutionary (e.g., Mikhail Bakunin's anarchism) or conciliatory (e.g., concepts of "rightful resistance").

As Lubis observed about his own life given to the modern peasant movement, land occupations are distinguished for their ability to provide material gains to the poor in the here and now, rather than waiting for the long-overdue rapturous revolution or slow-moving legal reform. For this reason, agrarian movements everywhere continue to be built around occupations. Reclaiming movements build new territories and along with them new kinds of self-determination, even emancipation. The domination and rule that seems to bind so tightly is revealed to be more vulnerable than previously imagined. Reclaimers expose the choke points. Ephemeral though it may be, direct action creates new forms of power in the popular imagination.

CHAPTER 5

Organizing the Movement

The real challenge is not just to understand or to prevent appropriation
of the commons, but to find the means to actually "steal it back."

—Jack Kloppenburg, 2010

With reclaimers' land occupation in the mid-1990s, the pendulum's swing
between smallholder and plantation control reached its apex in Casiavera.
Minangkabau smallholders had been dispossessed by colonial rule, their
children had reclaimed the land with revolutionary action during Indonesian independence, and then they had been dispossessed again as squatters
during the New Order. In the 1990s, a third and fourth generation of Casiavera's workers mobilized to reclaim the land once more.

Extending a lineage of dissent and activism, Casiavera's workers and customary leaders wrote letters, held protests outside the governor's office, and
blockaded the plantation road to shut down the plantation company. Then
the agriculturalists moved into a new phase of their reclaiming movement:
organizing it in such a way as to allow them to maintain their control of the
land and make productive use of it.

After a decade of experimentation in how best to organize the land,
reclaimers created a way of living with the land whereby the community (*nagari*) controlled the land in collective. Smallholders' use of the
land was mediated through the elected Community Council (Kerapatan
Adat Nagari), which defined how the land was managed and adjudicated
conflicts.

When the occupation of the land began in 1996, Indonesia's workers'
movements were inchoate, still in the stranglehold of the late New Order's
militarized intelligence state. It was a time of rural worker suffering and

industrial plantation expansion. But in late 1998 and 1999, after Major General Suharto resigned after street protests in Jakarta, an archipelago-wide surge of state and plantation land reclaiming ensued. As the New Order crumbled, landless peoples, smallholders, and Indigenous peoples occupied, blockaded, and destroyed the infrastructures and machines of capitalist exploitation projects. En masse tens of thousands of landless people, peasants, and Indigenous nations took to the land. In acts of protest, they destroyed the massive plantations belonging to state and corporate landlords. Agriculturalists planted their own selection of smallholder crops: coffee, rice, cassava, banana, rubber, oil palm, mahogany, and clove.[1]

Social movement networks quickly grew, once again becoming a global center of agrarian resistance, as they had been up until the regime's murders of leftists of all kinds began in 1965. In perhaps the most significant transformation of agrarian movements in Indonesia since the revolution, these reclaiming movements joined Indigenous peoples who had lost lands to the New Order development projects with poor agrarian and urban peoples who did not have any clear historical connections to Indigenous territories but wanted to reclaim the position of smallholder in Indonesian society.

Agrarian struggle involved collective action of many kinds. Collaboration between smallholder groups, Indigenous organizations, and environmental activists supported land occupations, protests outside of plantation company offices, and work slowdowns in the plantation estates. There were months-long sit-in hunger strikes at the presidential palace and urban protest marches in the cities. These planned actions were accompanied by more spontaneous acts of plantation arson and other sabotage, all joined with elaborate policy reform proposals and reports.[2]

Days after Suharto's resignation, representatives of fourteen rural workers' unions wrote the public charter for the precursor to the Indonesian Peasant Union Federation (Federasi Serikat Petani Indonesia, SPI).[3] Within the new, post-Suharto social freedom, the national agriculturalist movement for land, rights, and justice crystallized.[4]

Casiavera's activists seized the moment, so full of revolutionary potential, to solidify and organize their reclaiming.[5] As the New Order crumbled, the number of reclaimers occupying the land rose to some 250 families. Susilo, one of the youngest agriculturalists to join in this second wave of reclaiming, told me it was December 1998 when he went up to occupy the plantation "because Suharto had fallen." He joined his parents, after a long period of discussion, in walking through the company gates carrying machetes, hoes, water jugs, seeds, and saplings.

Neither Susilo nor his mother had previously enjoyed access to this land. Nevertheless, Susilo said he felt entitled to control of the land because he was "from there." Susilo and his parents had worked together to cultivate a level plot halfway up the land that is now, almost twenty years later, a diverse species, half-hectare agroforest. Freed from the deeply held traumas of the genocide of the late 1960s and the repressions of the New Order, Susilo said they lost their fear:

> We knew that our control was not yet certain, but we just went up and planted. From year one that has been our collective land (*tanah ulayat*)! Yes, there could be punishment! But why be afraid?
>
> We knew that during the New Order whoever was powerful (*berkuasa*) were the ones who could take (*garap*) land. But we planted the cinnamon [and other species of trees] to make our rights to that land clear. Before the land was empty. We knew the cinnamon could last into the future (*berselanjutkan*).

Susilo and the other reclaimers were not radicals. They were mothers, fathers, daughters, and sons, the majority of whom had labored on the Dona ranch and plantation and embraced the struggle to control the land and remake it into a smallholder agroecological landscape.

OCCUPATION ECOLOGY

Casiavera's reclaimers began their movement with action, a kind of guerrilla gardening. Loosely organized around the idea that anyone from the village could go up and use the land however they saw fit, a collection of women and men simply began clearing plots of grasses and experimenting with different forms cropping and livestock tending. To create the transformations they so desired, Susilo, along with his father, mother, and hundreds more reclaimers, eventually turned to the properties of trees—long lived, stable, resilient with relatively little care—to make their plots productive and establish their claims to the contested plantation land.

Susilo and his mother were fond of telling me about the permanence of the cinnamon and mahogany trees that the two had planted, giving them a defining advantage in their political project of land occupation. Indeed, over time "tree crops do something by their permanence."[6] As their trees grew taller and the trunks thickened, so did the durability of reclaimers' land claims because they represented the kind of land use relation that has long defined access rights in Minangkabau polities.

Tree planting was a durable investment of reclaimers' labor, proof that they had a long-standing engagement with their plot. The growing trees signaled to other smallholders in Casiavera and to the company that Susilo had a durable, economic relation with the land. If the company were ever to return and take control of the land, Susilo's mother would have been in a strong position to demand compensation for their forest farm, given tree crops' value. Alternatively, if their one-half hectare of planted forest lasted five years to come to maturity, Susilo and his mother could take their first harvest of cinnamon and clove, good for a few hundred dollars of income.

The agricultural forests that Susilo and the others cultivated were not the result of their land reclaiming; the forests were a central component of their reclaiming project itself. Cinnamon, mahogany, avocado, clove, and many, many more species became strategic participants in the struggle for Susilo and his mother to exert their rights to a piece of land and cultivate a smallholder livelihood upon it.

Susilo's experience shows tree planting to be strategic along two axes, territorial as well as economic. As an expression of territorial claim, it was without parallel. Planting annuals like chili or eggplant would not have secured the land in the same way, as these plants die after a few months of harvest and require constant maintenance, leaving no permanent mark of control across the landscape even while demanding constant tending. As for the labor the smallholders invested, tree crop commodities could result in the highest profits. Even building a home on the land would not have been as advantageous. It would have taken more labor and required a greater investment. In order to bring the full weight of a claim of homestead, Susilo would have had to live there, opening him up to potentially dangerous forms of retaliation for defying the company's state-endorsed claims in the uncertain, chaotic early days after the fall of Suharto's regime.

For Susilo, the economic dimension of planting an agroforest in the contested land may have been secondary but was certainly not inconsequential. If Susilo's main preoccupation was with a quick profit, grazing the land with a cow or two would have been the logical choice. Susilo's father had cattle at the time, the land was already under grass, and livestock were easily sold for cash on the volcano. However, there would have been none of the permanent territorial benefits of tree planting. Vegetable crops could have also returned a

FIGURE 9. Two contrasting plots on the reclaimed land. A still-uncultivated plot supports grass for cattle (left). A few hundred meters away, Susilo's diverse cinnamon forest grows (right).

quicker profit with good weather and diligent plowing, fertilizing, and weeding, but these time investments would have been risky given the political situation. Even a perennial like coffee could have provided a quicker return than the crops Susilo selected. But coffee in Casiavera brought along with it a colonial history of exploitation, needed careful tending to be productive, and would have had to be harvested continuously to be profitable.

Cinnamon and timber crops were the strategic choice. These tree species could take hold in the overgrazed, planted-out, and eroded soils if the initial planting was done properly, with little additional management required as the cultivars grew. These crops have the added benefit of being hardy survivors when confronted with grazing by the monkeys, boar, and other forest beings that lived just above the collective land in Aren's cloud forests and could eat an entire vegetable garden in minutes.

The trees supported Susilo and his family's property claims in the long run, and after some five years they provided profits equal to half the salary of a coolie laborer at Dona. Susilo's desire for a plot of land for enjoyable and profitable work joined with the ecological properties and material utility of tropical tree crop species to transform the landscape. By 2016 Susilo and his family had transformed the plot they cultivated into a diverse closed canopy agroforest, rich in the smells, sights, and sounds of tropical cloud forest life.

Yohanes, another man from Casiavera who went up onto the land upon the fall of Suharto, was not nearly as successful in transforming his family's reclaimed plot as Susilo. Yohanes knew the pace, tenor, and tone of a life as a menial laborer at the plantation better than most. He had worked there for most of his life before the reclaiming movement. When Suharto fell, Yohanes, his wife, and their school-aged children joined Susilo and the more than one hundred other families to go up to the land to plant their own crops. A bit of fear lingered from the past. Yet for a couple like Yohanes and his wife, landless laborers who had lost their only source of livelihood when the plantation shut down, reclaiming the land seemed to be better than the alternative: life as landless laborers, jettisoned into the "informal" economy of the nearby plantations or cities:

> We never knew when the company would come back. We knew the company wanted the land back. But I don't know anything about that, those problems. We just came up here, that was it. We came up [to live and work the land] because our leaders, the head of the Community Council and the head of the village, they said that we were going to share this land for the people. This was our land, from long before.

Yohanes never thought he was taking land that did not belong to him. He was reclaiming land that his community had lost to a corporate occupying force. He drew a continuity from the claims of his parents and their parents, based on ideas of being part of a community that considered this piece of the landscape to be rightfully theirs.

Unlike most of the reclaimers, Yohanes and his family did not have any other land, so they built a homestead on their new land claim. Of all the plots on the collective land, Yohanes's was the most symbolic because he and his family lived in the ruins of the Dona Company's ten-room office and barracks. It took me a few weeks to notice the ruin. A thick stand of clove trees and banana and sugar palms blocked the view of the old concrete company structure from the dirt road that divided Casiavera's land north to south.

Before the reclaiming, the Dona plantation manager (*bos*) and overseers (*mandur*) had stayed in the concrete office and barracks. In the late 1990s Monica Jessica Teuling had stayed there when the ginger plantation operation got started, scandalizing the religious sensibilities of many in Casiavera when she shared a room with her boyfriend. In 1998, just after Suharto fell, a group of young men from Casiavera demolished the plantation barracks. It was a self-evident statement. Yohanes told me, "It was one way we drove out (*mengusir*) the company." The ruins are a haunting mnemonic of a more

FIGURE 10. Reclaimers' home life in the ruins of the plantation office.

exploitative time. That they have been repurposed as a home for a family of smallholders makes them also a vivid elaboration of the work agricultural-ists have done with their struggle.

Most of the barracks is gone, but on its eastern side, four walls of the kitchen and the outhouse remain. Yohanes, a tiny and very thin man fond of wearing slacks, a wool-knit cap, and a fake gold watch, installed a wood roof over the kitchen and built a two-room wood home with a zinc roof on top of the wide poured-concrete floor. His family refurbished the surviving bar-racks bathroom, hanging a plastic tarp over its one missing wall for privacy. While I took in the symbolism of Yohanes's home in the ruins of the plan-tation manager's barracks, his son pointed to the tall chimney and fireplace built of bricks and plaster that mark where the office once was, now covered by the branches of a fifteen-year-old clove tree. "It was a very big room. Can you believe it even had a fireplace to sit in front of when it got cold at night?"[7]

On that first visit Yohanes was too embarrassed to invite me into his home built on the ruin. He and his wife lived in the home with one of their daughters and her husband, who were too poor to build a home of their own. When Yohanes did eventually invite me in, I found three grandchil-dren and one newborn great-grandchild napping under a single mosquito net. The family was living in a more severe form of poverty than most of those in Casiavera, who lived down below the collective land on their own family-lineage property. The family had no electricity for a TV, and there were no motorcycles parked outside their home. Their most important possessions were their *sabit* knives, used for cutting grass as fodder for their single cows. The well-worn and perfectly sharpened knives were neatly mounted up on the woven bamboo wall of the living room, out of reach of the toddlers.

As I sat with the couple to ask about the long years Yohanes had worked for the company and their lives now as reclaimers, neither Yohanes nor his wife made any broad political statements or declarations of freedom. In many ways living as a plantation coolie "was the same, like now." Yohanes told me. "We don't have anything. My son bought the new wood planks for our house, after the old ones rotted. I did not have any money in my pocket. I made the bamboo walls myself."

At sixty-five, Yohanes and his wife had not found prosperity on the collective land. Coming to agroforestry at a late age, Yohanes had struggled to learn how to cultivate a thriving agroecological smallholder farm. I ask him why he didn't plant more trees, like some of the others up on the collective land. His answer showed how much he still had to learn from the people around him: "Sugar palm, cinnamon, trees like that, they have long roots, so if you plant them nothing else will grow around them." The couple had tried planting chili, corn, ground nut, and tobacco, but never rice. They had a single papaya tree next to their home, along with a handful of older clove trees, banana and sugar palms, two avocado trees, and a few surian and petai trees. This home-side garden provided shade to the home but was not densely planted or expansive.

In the badly degraded soil, none of the crops did that well; the family told me they had never had a bounty crop to bring them the level of income that allows for savings and the trappings of a lower-middle-class life. They were just getting by. And just as he did when he lived as a laborer, Yohanes still spent much of his time cutting grass and tending his two cows, but now they were his own. His bananas provided a few dollars a week to string his family along between his small clove harvests. His family could "eat rice," a common refrain that alluded to Yohanes's precarious ability to buy rice for his family's three meals every day, rather than eat the yucca and corn that they grew themselves because they were too broke to buy rice.

When asked about what it meant to be able to live and work on the collective land, Yohanes became more emphatic as he recounted what the land

provided him with: a clean, tranquil, and safe place to live; friendly neighbors; and land for agriculture that provides a basic survival:

> Before we had no land at all. We were living with my parents, and my sisters
> could have told me to leave if they wanted to, but they didn't. We were not like
> some of the people who live down below [the collective land], with lots of land.
> I had no land; this is the place I found a life. I looked and got it here. I got a
> lift in life (*numpang*). There is no way I would have other land; how would
> I buy it?

Some twenty other families had joined Yohanes's to not only cultivate a
plot on the collective land but live on it as well. Their stories are varied, but
Yohanes's family was not alone in their economic struggles. Still, down to
the individual, they credited the land as a place where the poorest of the
community could live on a farm as a lifeline.

CREATING THE COLLECTIVE LAND

At first, after Suharto fell, Casiavera's Community Council only lightly
guided the reclaimers who came up to use the land. Even so, this reclaiming was not spontaneous; it was planned in discussion with the village head
of Casiavera and Daud, the mosque manager, peasant union member, and
former plantation laborer. Women and men from all the family lineages
spent long hours in deliberation. Not all agreed with the occupation, but
they did not object and allowed the direct action to move forward in a kind
of unstable consensus. All of these conversations unfolded within webs of
women-oriented ties to land, households, and family lineages that, to quote
Peggy Reeves Sanday, aim to create "peace," "suffuse life with joy," and "produce an aura of accommodation and cooperation."[8] As occupation numbers
increased, so did the coordination between the council and women's family lineages. The local ranking representative of the state (the village head)
and the highest customary authority (the council) coordinated to prevent
non-residents from claiming land, limit the size of each family's claims, and
mediate disagreements.

A few years later, in 2000, the council announced an effort to construct
a collective land management system. The land was to be divided into five
blocks. Those with already established plots would remain. The uppermost
block—still mostly unclaimed secondary forest—would be reserved for the
council to use to fund community infrastructure and events. The remaining

four blocks were to be managed by each of Casiavera's four hamlets. The council hoped that because these customary family-lineage leaders (*penghulu*) were already involved in many of their people's affairs, like weddings, funerals, and land transactions, they would be better at organizing the agriculturalists to make productive use of the land.

But even from within the Casiavera matriarchate established on mutual aid, passionate disagreements ensued. Reclaimers already using the land did not necessarily live in the hamlet that was designated to manage their plot on the collective land, creating confusion. Some reclaimers were concerned that the family leaders would use the blocks of collective land for their own profit. Others thought each of the blocks should be divided further, to individual families. Still others wanted to keep the land as a commons for grazing cattle, while those that had already planted vegetable and tree crops would not accept a commons with grazing cattle free to eat and trample their young plantings.

The reclaimers spent so much time discussing the terms of how each of the blocks would be managed that they barely ended up cultivating the land at all. They could only agree that those who had already gone up to reclaim the land to cultivate it should be able to continue to do so. A few people, mostly those living closest to the land in the highest reaches of the community, continued grazing cattle; planting, cutting, and selling elephant grass; tending small plots of chili, eggplant, and tobacco; and planting banana, clove, cacao, durian, surian, and mahogany. While ownership was collective, cultivation was an individual effort. Unlike the form of management that the community would eventually develop, in this early stage of reclaiming there were no formalized sanctions for disregarding the council's decisions about how the land should be used.[9]

The disagreements came to threaten the feasibility of the project of reclaiming itself. Overshadowing it all was the fact that by law the Dona Company continued to hold a valid state-issued land concession, foreclosing any possibility that Casiavera's local government and customary leaders would gain state recognition of their land claim as customary or indigenous land (*tanah ulayat*); the council were well aware that their ability to counter the plantation company's claim and hold onto the land depended upon agriculturalists' ongoing use of it. But whereas some reclaimers would have given up at this point, those in Casiavera did not.

In 2002 the council and village head began another round of discussions about changing the management of the collective land. Around this time reclaimer-led protests forced the head of the council to resign. Central to the people's grievances was the perceived mismanagement of the collective land.

Shortly after the community elected a new leader, the council wrote the first community regulation pertaining to the collective land. So began a long process of deliberation to put on paper a definition of acceptable use of the collective land and, above all, to reiterate the community's right to do so. With the full weight of the community's moral economy bearing down on them, the council were concerned about opening the process to full deliberation, so they wrote the regulation in closed-door meetings, with only the lineage leaders present. The resulting regulation reads, in part: "The ownership of the land will remain with the village. The land will be divided into one-quarter hectare plots. Every family that lives in the village will be eligible to use a plot. Land will not be bought, sold or rented."[10] Individuals with use rights (*hak pakai*) would cultivate defined plots. These plots were not to be inheritable. Tree crops could only be planted along the edges of the plots, to define their boundaries. Any homes built on the land were to be considered temporary. The elected Community Council of family-lineage leaders claimed the right to determine who had access to the land, create management rules, and adjudicate conflicts related to its use. If people using the land did not respect the regulations, the council reserved the right to repossess the plot and allocate it to another family.

With their first regulation, formalized in 2003, the council divided up the roughly seventy hectares of land into 264 rectangular plots of one-quarter hectare each. Building on centuries of sovereign rule of their *nagari* polities, the family-lineage leaders and council used their claims as a power of exclusion, declaring the collective land to be for the sole use of recognized resident families of Casiavera. Among the eight thousand or so Casiavera residents, access was exclusive but was not to be reserved for those of economic means. Entitlement was said to be based on need. Families without private wealth or existing landholdings were to receive preferential access.[11] The families who went up on the land and took it for their own in the early, freewheeling days of the reclaiming could keep what they had put under cultivation.

Within the year the council had allocated the remaining plots to willing reclaimers. Over the following years, hundreds of other people who wanted to claim a plot were not able to get one. Most difficult was the situation of the children of reclaimers who lived on the land, like Yohanes's six surviving children, who were too young to gain access to the collective land. Many of these young adults still lived up on the land and made long commutes down the volcano to work as construction workers, gasoline station attendants, and agricultural laborers. They had come of age too late to get plots of their own.

For the smallholders who did acquire plots, their rights to the land were contingent on their continuous use of it and the payment of a $2.50 yearly fee to the Community Council. If these conditions were not met, the land was to be returned to the community and "rotated" (*ganti gilir*) to the next person or cooperative waiting in line. A collector (*kolektor*) was tasked with collecting the yearly use fee and monitoring the use of the land. If people didn't use their plots, or used them in a way not consistent with council regulations, the collector was to inform the council, which could revoke permission to use the land or issue a fine or warning.

Susilo, one of the earliest and most active cultivators of the land, was elected *kolektor* by the council. At the time, Susilo was already an established figure of authority as the head of Casiavera's government-recognized People's Security Organization. Susilo began his work by identifying everyone who was using the land and recording their name and plot in a register.

Just as the authority of the village council was grounded in tradition, so too was creating the role of *kolektor*. The position of collector in Minangkabau began as a colonial custom, when Dutch administrators created the position as a way of enticing Indonesians to serve as colonial tax collectors and managers of the labor schemes of the Dutch's infrastructure and forced cultivation projects. In creating a collector, Casiavera's Community Council drew on entrenched Indigenous and colonial traditions. The council sought to control the land in a way that was local and collective, but did so in a way that retained forms of management familiar to long-existing structures of the colonial nation-state. It is possible the council's perceived need for a collector was a moment when their post-colonial imagination could not expand far enough to encompass non-colonial ways of enforcing and monitoring appropriate uses of the reclaimed land.

Casiavera's collector, Susilo, certainly presented an affect reminiscent of the way colonial collectors had extended their own power. This affect was underlined by Susilo's booming voice, buzz cut, broad shoulders, and liking for long black leather jackets—a classic tough guy persona all too common in Indonesia's plantation lands.[12]

Susilo's persona raises the well-known specter of the dangers of localism. Communitarian projects, especially when constructed to accomplish exclusions of outsiders, can be frighteningly authoritarian. There is nothing inherently egalitarian about local control.[13] However, Susilo's power was highly circumscribed by the council and the family-lineage leaders. When I was speaking with Casiavera's reclaimers, I found that the collector role was for most a second thought, the actual authority being the council, and it was the council to which people brought their problems and demands. Intimidating looks notwithstanding, no one ever spoke about Susilo as being threatening or violent, and I watched many times as Susilo cruised through the land on his motor scooter and casually chatted with folks in the coffee shops down below the land, in the village. He struck me as a kind, easy-going man with many diverse ties to the place, not only as collector but as a fellow tree crop cultivator, member of an organic vegetable cooperative, and small-time cow trader.

His look, role as collector, and being part of the People's Security Organization did bring up another question that is central to projects of collective self-determination: the need for unarmed movements to organize for self-defense and the enforcement of movement decisions. That the council felt the need for a collector with the image of an enforcer speaks to the fraught agrarian milieu that reclaiming unfolded within. When Casiavera's agriculturists made the decision to occupy the land, they rejected state law and entered Indonesia's vast realm of contested land politics, wherein violent toughs and land mafias are all too active. It is likely that people in Casiavera thought they needed someone with Susilo's persona on the land because of their reasonable concern that even more threatening toughs could show up at any time, attracted by the possibility of land or profits in a place where state authorities were absent.

LAND REGULATION CONTROVERSIES

As the drafters of the Community Council regulation of 2003 foresaw when they carried out their work behind closed doors, the implementation of their land regulation was not altogether smooth. The village head wanted to rotate the use of the plots every five years, so that more people in Casiavera would have a chance to access the land over time. The council instead fixed on a different consensus: married couples and cooperatives could hold non-inheritable rights to the plots for as long as they used them and paid an annual $5.00 tax to the council.

Daud, the manager of the Casiavera mosque, an active SPI member, and one of the first agriculturalists to claim land inside the plantation, recalled that this particular disagreement was unusually impassioned.[14] Illustrating reclaimers' fine line of survival, Daud's greatest concern was with the tax reclaimers would have to pay to the council for use of the collective land. The tax was unacceptable and unprecedented, just another way the powerful wanted to profit from the peasant. Daud spoke about how the tax touched a raw nerve that was connected to his family's history of tax revolts against the Dutch.

The council and village head sought a compromise with the dissenters. The head of the council explained that the taxes would ensure support from the wider community, from those who would not receive one of the 250 plots. Without the tax, the council head believed, support could evaporate unless everyone in the community felt that they were gaining a benefit from the collective land. Taxes would provide a benefit that could be spread to the entire community, through development projects like badly needed rice paddy irrigation and social events like concerts and soccer games. On this point the council head was insistent. Daud responded that he understood but that he and the others were still not happy about the compromise the council had offered, $2.50 a year for each quarter-hectare plot. But eventually they realized that what they were gaining in exchange was their local government and customary leaders' written endorsement of reclaimers' use of the company land, an invaluable position if the company or government returned in the future.

The most controversial part of the council's 2003 regulation was the ban on planting trees on the collective land. The council decreed that the land was to be used for annuals alone. But by then reclaimers' tree plantings were growing into diverse agroforests. The tree crop ban was to ensure the collective land would not eventually become private land. It was to be a public resource, its use rotated among local smallholder families over time. How could the land be a public resource to be rotated between users if rights to trees on top of the land were growing more durable by the day? With the 2003 regulation, the council attempted to impose their own tenure claims on top of that of agriculturalists because the council recognized what land tenure theorists have also come to acknowledge: that tree crops rearrange social relations as they take root and grow.[15] The council did not want to cede tree planting rights to the tree planters because they did not want the plots in the collective land to become the de facto property of the agriculturalists who cultivated them.

From the reclaimers' point of view, planting trees on their plots was first a way to contest the company's land claims, by providing both a statement of continued control of the land and an economic resource. In reality, reclaimers entirely ignored the prohibition on tree planting to cultivate cinnamon, clove, cacao, mahogany, surian, durian, avocado, and other fruit commodity crops. Reclaimers were compelled by the economic value of commodity tree crops, in the process creating a form of land use different from the one foreseen by the council.

As their trees grew, reclaimers came to speak about their trees as their individual property. With the trees and their incomes growing, women spoke about leaving their plots to their daughters. Mostly because of the limits on their own power within the community, the council were forced to let the perennial cultivation continue. The village head, on a walk through the collective land with me, told me that he had told everyone that planting trees was not allowed, but they did it anyway, and he did not feel he could do anything more than that. The sanctions against tree planting were not specified in the 2003 regulations, and he certainly was not going to dispossess his friends and neighbors after all they had accomplished.

Even more importantly, the village head told me, people using the land clearly did not support the ban. Plenty of family-lineage leaders did not support banning tree planting either, noting their own ancestors' long experiences with agroforestry. Yunus, an enthusiastic forest farmer and family lineage representative on Casiavera's Community Council, went so far as to develop a position that rejected any authority beyond the individual smallholder when it came to tree crops. "Avocados, rambutan, soursop . . . trees like these are not to be regulated by the state nor customary tradition (*adat*)," Yunus told me.[16]

Casiavera's Indonesian Peasant Union members were also unwilling to accept the tree crop ban. Tree crops and agroforestry were for them foundational to their ideas of how Casiavera's reclaimers would create well-being. The provincial chairman of the union, Arif, defended the union's support of planting tree crops in Casiavera with an appeal to pragmatism.

During a 2012 visit to the collective land with a group of union members from across Sumatra, Arif told the group that the agroforests surrounding them were the result of the people observing the facts and circumstances of this specific landscape, where they followed their daily, lived experience to determine the best way of using the land.[17] In the end, the council were unwilling to force the issue and risk exposing the limits of their power to control the collective land without the support of the reclaimers themselves.

After more than three decades of struggle, the collective land became for all intents and purposes managed by "the people" through the unique parallel support of reclaimers, a peasant union, a community council, and a village head. Despite the council and village head's close involvement, the land was never under their complete control, and reclaimers went about planting their trees.

When reclaiming became an effort of collective land control, it was not only contentious with the National Land Agency, the police, army, and administrative staff of the state; it was also contentious among the reclaimers themselves. Different ideas and perspectives about how to use the land came into competition. Yet the contention did not fracture the widely held idea that the land should be owned collectively, by the people who needed to use it most. After several iterations of cooperative management, reclaimers and their family-lineage leaders' collective control of the land held a place for smallholders to pursue their own ideas of well-being. Rather than risk the cohesion of their reclaiming with a rigid interpretation of how best to use the land, Casiavera's family-lineage leaders took a more flexible, pragmatic approach. At the most difficult reclaiming moments, like when reclaimers insisted on planting tree crops on the land and treating it as their own personal property, the council recognized the limits of their own authority and let the reclamation unfold as most reclaimers wanted it to.

Nurul was once his family-lineage's council representative. He told me the councils had lasting durability because of this pragmatism and consensus seeking. "It [council rule] is very similar to how it has been for a long time, of course it is flexible (*mulur*), but its core remains." Nurul told me that this core is a higher truth that guides social life, a truth held in what he calls the fire tradition (*adat api*), a custom that has come from the natural world, primordial and made all the more powerful by Casiavera's location on a volcano's flanks. It is a knowledge and set of practices that came far before Islam. Nurul described the *adat-api* in this way: "Everything comes from a seed. We are just at the start. We have a natural, an ecological scale. If we start a capitalist model, it will not match the earth, or the peasant. This not our way. We need consensus and discussion." If the council remained true to this knowledge and practice, they would remain successful, Nurul believed.

To finish our discussion Nurul told me, not for the first time, that the Minangkabau lineage councils are fundamentally *anarko*, or anarchist. The councils' authority is circumscribed. They operate through deliberation (*rapat*) and with consensus decision-making (*mufakat*) within parallel and mutually supporting women's and men's domains of decision-making. Male lineage representatives in the *nagari* community council could be replaced by their families at the direction of women elders, and they were if the lineage representatives began to exert authoritarian tendencies.[18]

There were a number of ways Casiavera's family lineages kept local authority in check. Council representatives' terms were not indefinite; council representatives typically served on the council for five to ten years. Throughout their term they needed support from all the elders of their respective family lineages, and the withholding of their support could doom a council representative's tenure.

During my time in Casiavera a few members of one of the family lineages went so far as to chain shut the doors of the customary council's chambers for three days, because in their view the council allowed their family lineage's council representative to represent them without their "approval" (*persetujuan*). The conflict was resolved when the chairman of the customary council said the council would only resume work if the council representative stepped down or resolved his conflict with the eighteen elders who had a problem with his actions. The council representative quickly consented to changing his position on a number of issues relating to political parties and use of local government funds, bringing his positions into line with the elders' consensus positions.[19]

For more than five centuries the councils have shifted and changed, their rules upheld and broken. Through it all they have continued to guide people's relations to each other, property, and the land. Even while the councils were incorporated and subordinated into legal structures of Dutch colonialism and the New Order, their representative function endured.

The Dutch inventions of the community head (Kepala Nagari) and community council (Kerapatan Adat Nagari) still exist, but that does not mean that the principles of deliberation and collective, matrifocal control of land have ended. Social power remains associated with the fertility of nature, a power that remains geared toward women's control of land and all the relations and commerce that unfold upon it. And while the patriarchies of the colonial and post-colonial orders remain dominant in state domains of power, within the Minangkabau territories men have no special rights derived from their gender. Only as brothers and uncles who act to protect (along with their sisters) rights to ancestral lands are men to have influence over the relations of production that unfold upon them.[20]

In Nurul's telling, the conception that a theoretical anarchism can be found in the council's long history began with Indonesia's greatest leftist theoretician, Tan Malaka (1897–1949), who was from the Minangkabau territories.[21] For Nurul and others in Casiavera, Tan Malaka's comment, "Where in Indonesia has the sovereignty of the lords been as restricted as in Minangkabau [West Sumatra]?" became a kind of organizing principle for their long land struggle.[22]

Writing from within a tightening colonial repression, Malaka was seeking a social form to replace colonialism. Malaka saw possibility in the Sumatran uplands and the norms of society there that limited the structures of a formalized nation-state as well as people's ability to exploit others. Yes, kings, sultans, and lords existed in the Minangkabau, but they were primarily charismatic or magical figures who did not have the power to rule other people, and they were unable to pattern daily life in society as a whole.[23]

In Tan Malaka's most famous work, *From Jail to Jail*, the author dreamed of a West Sumatran society in which communalism and mutual aid existed. For Malaka, it would be a nation where these anarchic principles could make life livable for most citizens. In customs of centering cooperative work, holding property in collective within a family lineage, and governing through a federation of matriarchal *nagari*, Malaka saw the foundations of a moral way of life. Part of this modern cultural politics was a reification of the past, of Tan Malaka's own peoples' pre-colonial history. Another part of this cultural politics was a recognition of land as a socially embedded resource, cooperative work as a way for people to prosper together, and mutual aid's underpinning of better chances of survival for the many.

In Casiavera, these egalitarian techniques of rule existed, at least in part, and were supported by a distinctive perspective on property. I often heard people in Casiavera who were concerned with their own past say that before the colonial encounter people "did not own anything." Family lineages controlled their own lands, while the customary council managed the more distant and marginal agricultural lands and upland forests as village commons.

According to Nurul, the moment that property became individual the agriculturalist became alienated from the *nagari* whole: "Property was the result of theft. Property is death." This statement does not mean that property is the death of human life, but that it is the death of the social system. In this perspective, property should be held in collective, to be passed through the current generation and onto the next unaltered, not fractured, forfeited, or accumulated. According to Nurul, it was this pre-colonial tendency to not alter collective property that allowed the Minangkabau to live for generations in *nagari* with stable boundaries.[24] Indeed, many Minangkabau intellectuals see the origin of their matriarchate in their ancestors' efforts to secure the home, food, and land the reproduction of their families required.[25]

Nurul sees the role of the Casiavera Council as central to their effort of reclaiming, saying that Casiavera's custom is "dense" (*padat*) as a way of conveying how well-developed and important it is.[26] Density means that Casiavera's agriculturalists have long nurtured a process of consensus rule that allowed for strong limits on leaders' powers.[27] The people who lived within Casiavera's boundaries imagined the land to be within their territory, belonging to them, to be managed by their council.

At the start of their reclaiming, the council recast the customary management of the land to encourage an overturning of the authority of the Suharto regime. They at once encouraged direct action, a kind of guerrilla gardening, and pursued a legal strategy through negotiations with the state agencies. This was a shift in leadership. The elected leaders, or *penghulu*, who made up the council moved from the side of the state to the side of the underclass to take a historically defiant stance. It had been more than one hundred years since the council had taken such a confrontational position. The shift was even more notable for the success the New Order and, before it, the Dutch, had achieved in co-opting the male lineage leaders, the *penghulu*.

From within an atmosphere of pervasive authoritarian repression, the Casiavera Council became increasingly convinced that their history gave them the moral justification to reclaim the land. Faced with New Order efforts to remake the council into one obedient to the regime and the ever-present chance of arrest or violence in retribution for open dissent, the council, following the feelings of the families they represented, moved from a place of acquiescence to challenging authority.[28]

As the council experimented with the exact form of collective control that would allow the reclamation to succeed, custom and cultural specificities provided the substrate on which they innovated. The council sought to effectively manage the collective land as community property. Their long tradition of political organization gave them the authority to do so. With this long tradition of authority, the council could claim the legitimacy to manage the land. The arrangement that would emerge was founded in custom but was a work in progress, an experimentation in systems of collective land control. Management of the land was not at all a fixed, unchanging practice that can be traced far back into the past of Casiavera's ancestors. It was instead a practice informed by custom that reclaimers and Casiavera's leaders considered to be a higher, natural truth that was crafted to match lived experience and immediate needs.

LAND CONTROL AGAINST DISPOSSESSION

After the 2003 regulation was finalized, community leadership marked out new quarter-hectare plots with a measuring tape. Reclaimers cultivated the land while Susilo filled his registration book with people's names and notes of their annual use-tax payments. Aren's workers went about remaking the plantation land into a landscape of high-value fruit, spice, and hardwood tree crops cultivated without tilling or chemical inputs. They sold most of their tree crop harvests to movement member traders, who went on to resell them in the Sumatran cities and for export.

Just below the cloud forests on Aren on the rapidly consolidating collective land, the concerns of the previous decade—state repression, the plantation, street marches, and blockades—seemed far away. But people's thoughts often returned to the fact that the state or the company could dispossess them from the land yet again, "from above." Reclaimers' concerns led them to create individual plots on the collective land with the hope, which turned out to be correct, that their full use of the land would make its productivity obvious to the government, which would then be hard pressed to evict them and return the land to corporate control.

Casiavera's reclaimers did not only have to worry about such land dispossessions from above. They also had to worry about a second kind of dispossession, in which the most affluent and influential smallholders would find a way to accumulate plots in the collective land from their neighbors as time

TABLE I   CHANGING LANDHOLDINGS ON CASIAVERA'S COLLECTIVE
LAND, 2011 TO 2016

Plot Characteristics	2011	2016
Reclaimed by families	194	141
Reclaimed by agricultural cooperatives	8	6
Average plot size	0.42 ha	0.31 ha

Source: National Land Agency's 2011 district inventory survey of Casiavera. I created the 2016 numbers from home and field visits and by consulting the community secretary's land records.

TABLE 2   PLOT TURNOVER AMONG RECLAIMERS, 2011 TO 2016

Reclaimer	Maintained Plots	Lost Plots	Gained Plots
Families (n = 232)	100	90	42
Agricultural cooperatives (n = 10)	3	5	2

Source: National Land Agency's 2011 district inventory survey of Casiavera. I created the 2016 numbers from home and field visits and by consulting the community secretary's land records.

went on: dispossession "from below." To limit this internal depeasantiza-tion, the council decided to prohibit any ownership or transfer of the plots among reclaimers. The total area open to each smallholder, one-quarter hectare, was limited to a holding that would only provide a basic livelihood. Withholding the land from the market from both below and above, the collective land became an asset of counter-dispossession that was meant to ensure the ability of the poor and marginalized to access land.

A closer look at how the land was used over time offers insight into the way the limits on dispossession from below have played out thus far. One way to do this is by comparing the land use registry the National Land Agency prepared in 2011 with my observations about how the land was being used in 2016 (see table 1). In 2011 the office recorded 194 families working plots. After carrying out a survey of the land and consulting with the village secretary's records in 2016, I found that not all of these smallholders who had gained access to the land had found success.[29] Some 90 families and five cooperatives had given up their plots over a five-year span (see table 2).

These families who failed to create a smallholder farm on the collective land could not overcome the ruined soils on the old plantation or were overwhelmed by the labor demands, skill, and luck needed to profit from the land. Most of the ninety families who stopped using the collective land cultivated plots of one-half hectare or more, while at least forty-two new families had received one-quarter hectare. Shrinking plot sizes indicate that the agroecological challenges of rehabilitating the ruined soils of the collective land are very real. The shifts in the number of farmers and plot size also suggest that unsuccessful reclaimers stumbled against the kinds of challenges that contract farming, sharecropping, the weather, debt, and the lack of agricultural expertise present. Still, some two hundred families were cultivating the land. Determined, experienced, and fortunate small-holders persevered.

While the council declared that land access was to be based on need and order of request, the land use registries also show that kinship ties and community standing influenced who gained access to the land. Influential individuals used their positions to convince collector Susilo to issue plots adjoining their own to their siblings, even though they intended to control the plots themselves and cultivate them with waged labor. In this way a few individuals accumulated access to plots in the collective land greater than the permitted one-quarter hectare. But only two people gained control of one hectare or more of the collective land. The first was Zain, the lawyer from Casiavera, and the second was a male lineage leader with a title that carried the highest customary authority, Datuk Paduko Marajo.

Taken together, these numbers suggest that the collective management of the land has throttled back the dispossession of smallholders from it, but not completely. Most importantly, these numbers do not show any large-scale accumulation of plots in the collective land over time.

With their combined collective control, individual cultivation system, Casiavera's land managers attempted to strike a delicate balance. The system allowed individuals to establish commodity crop production for personal gain. At the same time, the land is inalienable by powerful or wealthy individuals and corporations. The allocation of non-inheritable and non-transferable individual use rights was formalized with Casiavera's creation of clearly defined individual plots within the collective land, maintenance of a use registry, and the active redistribution of unused plots to residents without land access.

In confronting land dispossession, Casiavera held a strategic advantage: many in the community already possessed deep knowledge of capitalist dispossession. More than one hundred years before the New Order's Dona Company, the colonial enclosure movement had brought dispossession from above to the volcano with Dutch colonial forced coffee cultivation and W. H. Samuel's logging and cattle ranch. Even before these colonial dispossessions came hundreds of years of dispossession from below associated with Sumatran smallholder commodity economies centered on valuable crops like coffee, cinnamon, clove, durian, rattan, resins, and tobacco, whereby fortunate cultivators, collectors, and traders accumulated wealth and farmland from their neighbors, driving class differentiation.

The long encounters with commodity crops taught Casiavera's reclaimers that unless they constructed and maintained some form of land control to counter dispossession, the multifold pressures of agrarian capitalism would eventually dispossess them from their lands. In organizing the control of their land, smallholders, customary leaders, and local government representatives worked together to reconstitute a form of land control known for hundreds of years in the region.[30] But this land control knowledge is not a holdover from pre-capitalist Indigenous traditions. This knowledge co-emerged with commodity economies and, eventually, capitalism.[31] It was a generations-old practice of collective land management as response to capitalist dispossession. Reclaimers innovated on this lineage of land control to build up the social mores required to confront dispossession. More specifically, Casiavera's Community Council used their institutionalized knowledge of collective land control to design and codify a form of land management that throttled back the dual social pressures of capitalism and smallholder commodification.

The existence and capacity of institutions to protect smallholder well-being cannot be assumed to exist. Moral economies must be created, foremost from a knowledge that without them smallholder economies will be fractured both from above, by state and corporate land grabs, and from below by wealthy and powerful individual smallholders. Like all social struggles, the authority and ability to continue these counter-dispossessions' relations to the land required continuous negotiation and effort.

During my time there, Casiavera's customary council, landowners, and users of the collective land were constantly in discussions about the land and its use. Meetings of family-lineage leaders in their various homes about their family's private land matters happened nearly every week. A marriage, a proposal to lease land to outsiders, a request to sell a piece of land to pay off debts, an idea to grow root crops on a plot of jointly claimed land to sell to a nearby military base mess hall—all were reasons to meet and discuss. It was up to the family lineage to guide the discussion toward consensus.

Discussions about the collective land were every bit as common. Even nearly two decades into reclaiming, serious disagreements between people in Casiavera about the most fundamental attributes of the collective land existed, for example, if anyone from the Community Council would challenge the reclaimers' ubiquitous tree planting and if the land would be inheritable. Even disagreements about the boundaries of the collective land continued. This last issue was likely even more controversial than the prohibitions on tree crops and inheritance.

After a new village mayor (Wali Nagari) won election in 2014, a few members of the customary council decided to try to expand the boundaries of what the village government considered collective land. Up to that point, reclaiming decisions had been based on the idea that a road and fence that the colonial-era logging company had built defined the lower boundary of the old land concession and the now collective land. Five or six households had for at least three generations lived on the strip land directly on the downslope side of this road and fence. In May 2015 council members requested council approval for a new regulation that would require these families to pay the collective land use tax that everyone living on the collective land started paying in 2003. According to these council members, the households were not on private family-lineage land at all; they were on the collective land.

A series of meetings took place to visit the disputed land and talk to the households caught up in the land affair. Because it was a land matter, women lineage leaders from all Casiavera's recognized families attended, along with their uncle or brother counterparts, who would argue their family-lineage positions before the Community Council. The first meeting on the land included some fifty lineage leaders and council representatives.

Daud, the mosque manager and SPI member, lived in one of the homes on the contested land. He consulted with his SPI friends and then wrote a letter to the customary council. He wasn't hopeful, though. Like others in SPI in the community, Daud felt that the push to expand the collective land was coming from the newly elected mayor, who was said to be looking for investors to bring in a developer to build a hotel on the land in question. In his typically sardonic way, Daud told me the real issue was that "we don't continue to organize the way we did before [during the land occupation]. When we don't organize, we are flimsy (*tipis*) when it comes to these issues with our own leaders."[32] Another SPI member was even more upset and went to the provincial SPI chairman to lay down plans for a protest if the council went any further. The plans were not needed that spring, however.

During a meeting on the land in April 2016, it was clear that the council's initiative to expand the collective land had failed. After speaking with many who were "contra" (*kontra*), a number of council representatives made their objections known. One said to the group, "I am contra! I have told my family that nothing will change. I know the history; I know the names [of the people who have lived on this land]! The *wali nagari* has been off in the city, he does not know the people here or those that live on the collective land."[33] Several of council members nodded and spoke their agreement. A few minutes later the council members began to walk back to their waiting cars at the collective land entrance gate, speaking quietly in groups of three or four.

Five days after the meeting, another village elder told me, "They wanted a project on the land. If so, there was a protest coming. Where will our people gain a livelihood? This land is to make us prosperous (*makmur*). This [changing the land boundary], for us, is like burning us out (*hangus habis kami*)!" It seems the contra sentiment was widespread. A trader, smallholder cooperative member, and SPI organizer told me it wasn't even the idea of the new tax on the households that was the problem, but that private property would be lost to the government and an outside investor would be brought in: "We just got this land back from the company. I will not let people become laborers on their own land again. Rules, taxes . . . fine. But I don't agree with [bringing in] an investor."[34]

THE LEGAL CHALLENGE TO CASIAVERA'S RECLAIMING

Although reclaimers had not seen Dona staff on the land for years, they were correct to worry about the return of company control and the possibility of another dispossession from above. As it turned out, Dona's owners had very much kept Casiavera and the reclaimers working the land in mind. In March 2012 a challenge to smallholder control of the land appeared when the holder of the Dona Company land concession permit, Monica Jessica Teuling, brought a civil lawsuit against Casiavera's customary leaders and village head. A reading of reclaimers' and company staff's testimony along with the final decision of three West Sumatran judges explains much about the ways that corporate control is brought to bear on smallholder agriculture, as well as the ways workers can take up the law to support their reclaiming of plantation spaces.[35]

In her complaint, Monica Jessica Teuling alleged that a group of men from Casiavera had "authorized unscrupulous people" to take control of the plantation land, with the effect of interrupting operations on the plantation and ending Dona's business. Dona's legal team labeled the defendants "rogue officials" (*oknum*) that had caused economic damage to Dona. Teuling's legal team alleged that the reclaiming cost the company millions of dollars. The company's lawyers claimed Casiavera's occupation blocked the company's 2011 plan to establish a "sorghum plantation" operation in partnership with a mid-level Indonesian agribusiness, causing further financial losses. The plaintiffs sought the return of the land and financial compensation. Their complaint was overtly written as an effort to bankrupt Casiavera's leadership and regain control of the land by judicial order.

Teuling relied on an Indonesian legal argument well-known in the law, namely that because the land was once a colonial-era land concession, an *erfpacht*, the state acted legally to lease the land to Dona. Teuling argued that since 1970 the company had operated a profitable livestock and plantation business on the land. Teuling claimed that the company had invested in the community and infrastructure, including a road and irrigation projects, and that from 1995 the land was under large-scale ginger production. According to company staff testimony, the company operated without incident for a full three decades. Dona was said to "provide jobs that continually increased the local economy." Dona staff represented their intervention as benevolent and without objection, and so staff were confused when Casiavera's laborers became "squatters" and "looters," as the plaintiffs labeled residents. Thus, the unilateral actions of Casiavera's leadership, who had encouraged the "occupation" of the land, were unreasonable. In their closing argument, Dona's lawyers wrote that the judges should order Casiavera's leadership to pay damages and return the land, "with aid from the tools of state power."[36]

A team of lawyers from the Indonesian Legal Aid Foundation (LBH) and the Peoples' Coalition for Justice, Democracy, and Human Rights (QBar) represented the defendants. Casiavera's reclaimers' connections with these organizations had been founded a decade earlier, when Casiavera's leadership sought legal counsel at the start of their occupation. With deep experience working in Casiavera, LBH and QBar constructed a bold legal argument notable for the precedent it could establish for smallholders. Representatives

of the Casiavera Council and seven other smallholders using the land testified for the defense.[37] They all acknowledged reclaiming the plantation
land. This was incontrovertible; the three judges ruling on the case would
see so for themselves during a one-day field visit to Aren in late 2012. What
the defendants chose to argue was the significance of this reclaiming. Was
Casiavera's reclaiming a violation of Dona's business rights? Or should it be
considered a reasonable step taken by rural workers to rectify the wrongs of
a history of dispossession?

To begin, testimony and argument stressed that Casiavera had a history
of using the land. Smallholders had worked in collective to manage it before
they lost the land to Dona in 1970. During that time no one in Casiavera
had given their free consent to the company's control. There was no purchase
payment, called a *siliah jariah*. The defendants concluded that the land was
simply illegally taken. Compounding the injury of the loss of the land was
that no one in Casiavera experienced any improvement in their lives when
the plaintiff, Dona, controlled the land. Casiavera's reclaiming made the
land more productive than Dona ever had, the defendants testified. It was
a productivity that supported their own well-being. The defendant's legal
team underlined this point again and again, claiming that more than two
hundred families and twenty-five smallholder cooperatives were cultivating
the land at the time of the lawsuit.

One of the village heads of Casiavera during reclaiming, Riza, was a
named defendant in the case. He testified that Casiavera's reclaimers were
using the land for what he called "commercial agriculture." Riza said he had
worked with the council to create the 2003 management regulations of the
collective land, to bring "clarity and certainty to Casiavera's collective land."
Riza joined every single other witness that the defendants called in claiming
that he had no knowledge of Dona's operations or legal standing. This was
untrue. The village head and others knew that Dona operated a ranch and
plantation, where Casiavera's people worked as waged laborers. They were
also aware that they had reclaimed it from the company. Their denial was
subterfuge; by law Dona had to operate with certain standards of transparency with Casiavera, and by claiming no knowledge at all of Dona, Riza
could paint the company as being in violation of these laws.

Weighing the arguments, the court began its decision by summarizing
the evidence it took to be fact. First, Casiavera had controlled the land since
at least 1950. This lineage of control was strengthened by the 2003 regulation

concerning management of the collective land. Although Casiavera's use of the land was illegal, the court cited customary law to recast the reclaimers from squatters to smallholders and Dona from victim to exploiter. The judges wrote:

> Casiavera society was dominated for generations. . . . It is untrue that the defendants committed any appropriation of the land, but only carried out their right to manage the land for their increased welfare. . . . It is clear that it is the Plaintiff who has committed an unlawful act in this case.

The judges took issue with plaintiff Teuling's spin, not least that "the plaintiffs claim that they have never been contested for thirty years." The judges demonstrated the plaintiff's lie; the court documented a long oral and written record of dissent in Casiavera about the company lodged at all levels of government. The judges also noted that the state previously had recognized the reclaimers by building a road on the erstwhile plantation, to help the smallholders traverse it and bring their agricultural harvests to market below the volcano. The state also had built irrigation works, as well as a limited piped-water system, for drinking water. In the view of the court, all these government infrastructure projects contributed to the legitimacy of the reclaiming.

The court also addressed Dona's legal standing to operate. It found "errors in objects of the 1970 HGU, extended with HGU No.1 1997." The court found that in these HGU documents the land concession boundaries were not specified correctly. The boundaries did not have any geographic meaning that corresponded to Aren topography. Evidently, in 1970 Mahmud Teuling and his staff had simply counted off an area of seventy hectares and built a fence. There was a second, equally troubling problem with the HGU. The court found that the 1970 HGU had been issued without the permission or knowledge of Casiavera leadership or the community as a whole. The judges cited a 1977 Supreme Court decision that required the clear and complete communication of HGU land boundaries as part of the land lease process.

Reflecting the changing expectations of Indonesia as a democratic state, the judges admonished Dona for their secrecy and opacity as they worked in Casiavera. The result had been "forced evictions" of Casiavera smallholders from their "hereditary" lands. It was the plaintiffs who had "usurped, taken and mastered it [the land] without right." It was an action against Casiavera's "communal land law" without the permission of the community. As for the question of productivity, the judges' own observations of the land during field checks for the case led them to rule that Dona, "did not properly implement development" of the land. It appeared to the judges that Dona had abandoned the locale completely.

The court's rejection of Dona's standing to act as "master" of the land could have established an important precedent for smallholder rights and the validity of customary law. But in the end, the ruling preserved Dona's HGU and did not categorically improve the cultivators' position in the long run. The judges took great care to only rule on the plaintiff's lack of standing to bring the case. The court gave no ruling on who was legally allowed to control the land, avoiding endorsement of Casiavera's reclaimers or the company. The court decided only that the named defendants were not shown to have participated directly in the reclaiming and so there was not any proof that they had caused any damage to the company. In this limited ruling, the court did not rule on the validity of the company's grievance itself.

So, while Monica Jessica Teuling and her legal team failed to win a ruling that would help their effort to control the land, the ruling also did not provide any legal certainty to Casiavera's reclaimers. They were left with their guerrilla gardening, their direct action, and their reclamation. And making sure the land was under cultivation became even more important, as the news spread that Dona's right to an HGU had survived the lawsuit intact.

The importance reclaimers placed on the use of the land struck me when one day I made the mistake of telling Tia, who was working her cacao trees, that there was a wide area of empty land next to her orchard. "No! That is not empty, that is grass that we planted. We planted that grass to feed to our cows. It was hard work planting that grass. There is not any empty land here. It is all used." Tia continued this way, staring at my notebook, making sure I wrote it down. Her concern was that it should be clear to me, and to whomever I might pass on this information to, that the land was being worked. That it was being improved and utilized. If the land was going empty, it could be used as evidence by Dona's management and lawyers for why they should be allowed to return to control it. If word got out that there were unused areas in the land concession, perhaps the company would be able to convince the National Land Agency and the police to evict Casiavera's reclaimers.

Reclaimers built their movement around the idea found in all common law systems that landownership derives from use and dwelling. With such awareness, reclaiming leaders worked to make sure not only that much of the land was cultivated, but also that more than twenty families were able to build homes on the collective land. The reclaimers cleared front yards, planted flowers and trees, ran electricity wires from their neighbors, and put up television satellite dishes—all the better to make material their reclaiming and demonstrate their connection to the land.

Only the National Land Agency's cancellation of Dona's land concession permit would alleviate reclaimers' anxiety that according to the law, the company could still return to take control of their land any day until September 24, 2027, when the company's land concession permit will expire.[38] It was a heavy worry to carry, but not overwhelming because reclaimers knew that the law did not fully determine life on the volcano, nor had it ever done so. A shared perception of their dispossession as both unjust and changeable motivated reclaimers' refusal to allow laws to determine their lives.

Casiavera's reclaimers were bound together by their creation and practice of land control. As the erstwhile plantation became a smallholder landscape, it became a source of liberation from the oppression of plantation overseers and the violent fluctuations of corporate capitalism. New social relations of cooperation were built, like those on display among SPI members, cultivators of the collective land, and private landholders who worked to defeat the new mayor's plans for outside investment. When reclaiming became recommoning, the land became the site of a resource used for social benefit, a buffer from the larger problems of the countryside, and a step toward an economy that supported cooperation and equality. As an ongoing experiment in organizational and property politics, Casiavera's reclaiming suggests that anti-establishment, loosely organized efforts of collective control are indeed capable of transforming landscapes and sustaining workers' livelihoods.

CHAPTER 6

Diversifying the Land, 1998–2016

"The company made the land barren, with their tractors and their chemicals," Wangga—one of the most enthusiastic reclaimers in Casiavera—told me as we stood in her plot on the collective land. That day Wangga was dressed for work. She wore shin-high rubber boots and held a curved fodder knife. Her simple brown headscarf was tucked into a long-sleeved T-shirt that had on it the photo and campaign slogan of a local politician who had visited Casiavera a few years back, handing out shirts while looking for votes. I took to spending time with Wangga because over the years she became known in Casiavera for her skill and success working her plot.

After working as an office janitor and selling fried rice from a roadside stand in the provincial capital, Wangga, a mother of two teenage boys, returned to Casiavera to transform a badly eroded plot on the collective land into a productive agroforest. When Wangga began to cultivate her own plot on the collective land in 2000, it was, according to her, "an extreme waste ground" (*paling gersang*). Generations of industrial abuse left the plantation land badly degraded. Just a few patches of the tenacious *alang-alang* grass and a few shrubs grew in the compacted and salinized soil.

After W. H. Samuel and Dona Company had cleared the forest and destroyed the soil, Wangga went about rebuilding it. Others had tried before her to reclaim the same plot but gave up after their crops did not take hold. "They told me it would not work, planting here. They said nothing will grow, not even grass!"[1] It took a certain mix of need and determination to see potential in such a wasted place. While ruining the land was for a few decades a lucrative business, reclaiming it seemed to many a fool's errand. Even as some gave up their reclaiming, others persevered to harness ecological processes to spark the restoration of the land.

Wangga's plot on the land abutted a steep ravine. A still uncultivated plot bordering hers was covered in *alang-alang*, a grass that thrives on fire-prone, eroded, and exposed soils. A living fence of twelve avocado trees, all thicker around than her shoulders, followed the plot boundaries.

A massive candlenut tree (*kemiri*), fifty feet tall, grew at the southern edge of her plot. The candlenut was the first tree she planted; it grew from a seed her sister gave her from her own agroforest at the foot of the volcano. After sprouting the seed at her house, Wangga carried the sapling up to her plot to plant it. Around the same time, Wangga started keeping chickens on the land in a cage that her husband built from bamboo. When she had saved a bit of money, she bought a goat. After a year or so she made enough to buy a cow. With her cow she had a source of manure to improve the soil ("I carried the manure around my plot myself, on my head!"). She planted bananas to give shade to the cacao she planted next. Her neighbors gave her the banana saplings and cacao seeds. She planted the cacao sapling by sapling, seeing to it that each one took hold before planting more. She learned from friends how to tend the plants.

On the north and south sides of the plot was a line of mahogany and surian. She planted these trees to have the boundaries of her plot "certain' (*tentu*), so that everyone would recognize this land belonged to her. The far side of the plot transitioned to elephant grass, for fodder. Alongside the candlenut tree was a water tank that Wangga dug and lined with a tarp, to use to water her cows. A few feet beyond that was the cow pen that her husband built from small timbers and bamboo. She added a roof of grass, to keep the cows cool. Surrounding the pen was a three-story agroforest of banana, clove, and cacao. Mixed throughout were a few durian and other fruit trees, betel nut, avocado, chili, and many more cultivars, all of which Wangga planted herself over the past fifteen years. Her clove, now just a few years old, came to harvest for the first time in 2015, yielding half a kilogram of the aromatic flower buds for the market.

At the edge of Wangga's lively agroforest the bordering plot just to the north lay still unplanted, abandoned. Here the industrial monoculture ruin of the Dona plantation agriculture and Wangga's reclaiming were clearly

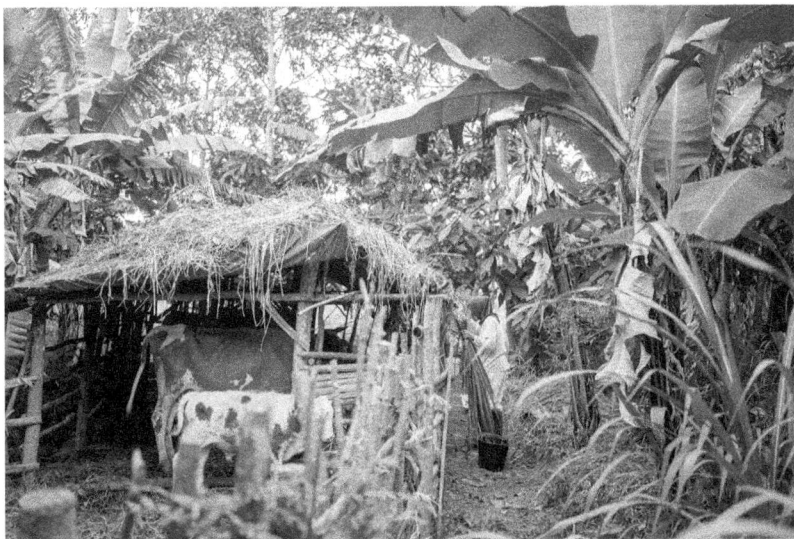

FIGURE 11. Wangga at work in her plot.

contrasted. They are two visions of agrarian change, development, and nature, inscribed into two bordering plots of land.

As I came to know Wangga better, I realized that like nearly everyone in Casiavera, she had lived a life of mobility and waged labor before she came to seek a life as a smallholder. After high school, Wangga tried working as a laborer in the Dona plantation, then moved to the city looking for a better way of life. Her closest sister, Diah, remained at their parents' home in Casiavera. As the eldest daughter, Diah was the single heir to Wangga's parents' small rice terraces, home garden, and homestead. Diah married a man from the other side of the volcano, who following custom came to live with her. Wangga told me that her sister would have let her stay, all those years ago when Wangga, still a teenager, decided to leave home. But Wangga said that the home and its land would not have been enough to support her hopes for livelihood and a family in addition to Diah's.

Instead of staying put, Wangga spent a decade in the busy port towns on Sumatra's west coast. She married there, to a fisherman from a town "where there were not any fish anymore." The two of them learned together how to sell fried rice on the roadside, clean offices, work construction, and do a bit of petty trading. When Casiavera's residents reclaimed the land above the community, Diah sent word to Wangga that there was land to be claimed for those willing to work it.[2] And so Wangga and her husband returned to reinvent themselves yet again, as smallholders.

Walking back from her plot, Wangga and I noticed a few women taking shelter from the afternoon sun in a grove of fruit trees—a women's work exchange group laboring for the day harvesting one of the women's rice paddies. They invited us to join their lunch of jackfruit curry, chili, and rice, and we sat. As my eyes traveled across the landscape around me, I asked about the diversity of trees planted around us and the way the land had changed. Wangga spoke first. "It's not like with your giant farms in the USA,' she told me. "Here we have many species. See, sometimes the price of clove rises, or the cacao falls, so we have to have many crops." She pointed up the slope of the volcano to her small field on the collective land and listed the tree species she herself had planted: "bitter bean, cacao, betel, sugar palm, avocado, durian, candlenut." She mentioned pepper, but I could not see it growing. "Where is that one?" I asked. Wangga laughed, pointing. "It is right there! It is climbing that tree there, the rain tree (*kayu hujan*). We plant them together. It is the tree that waters the pepper. It collects the water with its leaves and gives it to the pepper. We always plant it first, so the pepper will climb off the ground."[3]

As Wangga and the other reclaimers manipulated plants across time and space in a series of intimate, human-scale interactions, reclaimers' reconstituted agroecological knowledge deepened.

FOREST GENESIS

Wangga's cultivation of the damaged landscape was not a spontaneous process of restoration. She provided the conditions that led to the return of ecological function and the eventual emergence of a forested landscape. To do so required enhancing the damaged soils and increasing species diversity, biomass, and productivity. Nearly a century of logging, grazing, and monoculture planting eroded and compacted the soil's once rich layer of humus, created over millions of years of volcanic activity and forest decomposition. To support new life the soil required an input of minerals, nutrients, seeds, and work.

To begin, the reclaimers burned their plots, killing the grass and shrubs and providing a boost of nutrients to the land. Most chose to start in

Sumatra's late dry season (May–August) to reduce the nutrient runoff and compaction that the rains can cause. Reclaimers dug wide holes in the ruined soil to plant their banana and sugar palms in a starter soil of food compost, manure, and grass cuttings. Banana palms were a popular choice to start planting because the palm flourished with direct sun and created its own shade for other plants below it. As the banana palms took hold, the reclaimers scattered seeds of chilis, eggplant, peanut, and tomato. After a few months, once the banana palms had grown head high, they harvested their vegetables and planted tree saplings in the palm's shade, like cinnamon, clove, avocado, and mahogany, among some thirty additional cultivars. The smallholders brought seeds and saplings propagated in their homeside nurseries. Neighbors shared seeds from their home gardens. They made their own fertilizer, a compost of cow manure, dried rice stalks, and ash.

Great care was needed to establish their young saplings. Reclaimers learned that mounded-row planting worked best to start their trees, to alleviate soil compaction. When the plot was open to full sun, judicious weeding was required to ensure the survival of plantings. After a year or two, the reclaimers were able to reduce the amount of labor they put into their plots, when human "management" became secondary to plant physiology and plot ecology.

After reclaimers selected, planted, and weeded their tree saplings, their agroforests grew through interactions of light, oxygen, water, soil, the microbiome, and pollinators. A reciprocal interaction between soil and trees began. The new plantings created the microclimate needed for their own survival: shade, moisture, and fruit to attract seed dispersers. Insects, amphibians, reptiles, birds, and mammals moved in. Along the ground story lianas and grasses spread. Above these were trees that need shade to flourish, like coffee and cacao. Above it all were the fruit and timber trees, growing into a canopy as they sought the sun. The canopy cooled the soil and blocked it from erosion from heavy rains. Falling leaves, branches, and faunal remains accumulated and were eaten by decomposer bacteria and fungi to create a top layer of rich, spongy humus. The canopy's shade in turn prevented evaporation out of soils and streams and captured moisture out of the air, dripping it down into the understory. As years passed the canopy closed and the annuals phased out. Nitrogen-fixing trees and mineral-emitting rock maintained nutrient balance.[4]

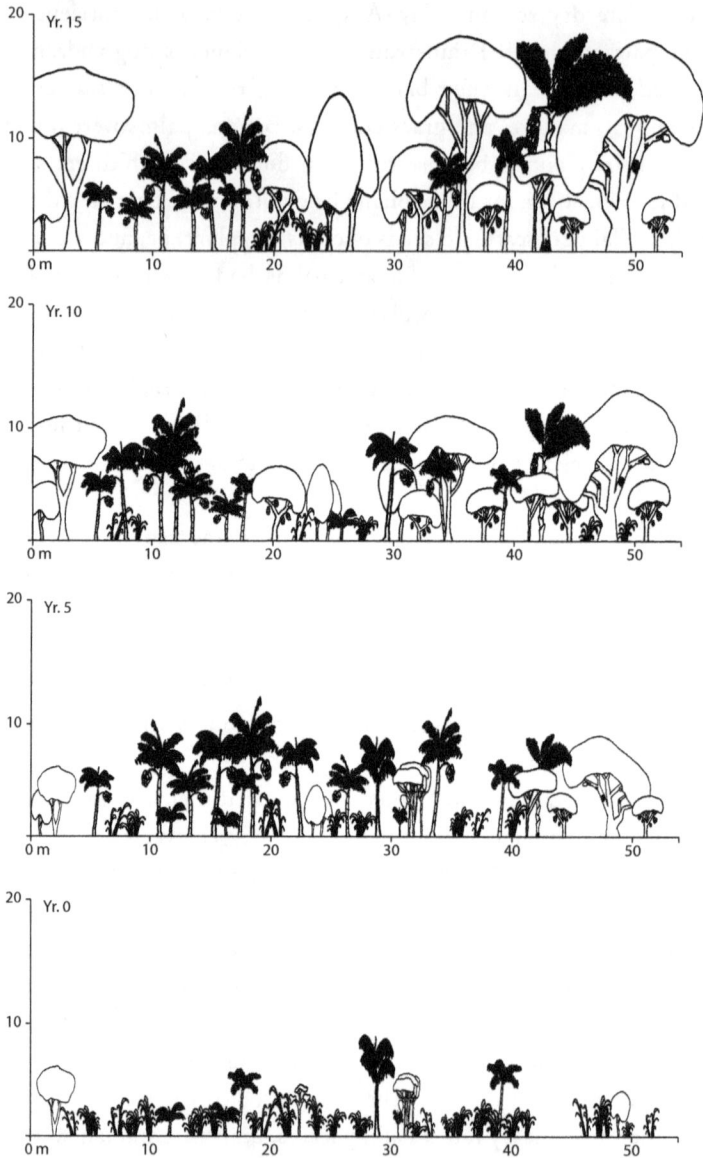

FIGURE 12. Drawings showing a mixed clove, cacao, and banana agroforest growing over time (years 0–15). Year 0: Planting of banana, clove, and cacao along with
other tree and palm species. Year 5: Banana and grass are main elements of agroforest composition. Clove and cacao become productive. Year 10: Clove, cacao,
candlenut, avocado, and sugar palm begin to form a forest, shading out banana
and grass. Sugar palm becomes productive. Year 15: Avocado, kemiri, and sugar
palm establish an emergent canopy. Cacao and banana continue as shade species.

FOREST EMERGENCE

It is useful now to take a step back from Wangga's and other reclaimers' plots to examine the surprising socio-natural emergence of Casiavera's reclaiming agroforests across the landscape. To do so I take up discussion of satellite photography, because it allows me to traverse time and the multiscalar political ecology of the Aren volcano, linking my work with individual reclaimers to their propagation of hundreds of now-forested plots.[5]

A collection of satellite photographs that spans twenty-six years of change across the collective land allows appreciation of how ecological diversity and forest cover increased across scales from individual plants in reclaimers' plots to a landscape measured in the hundreds of hectares. The first image, from the Landsat satellite, illustrates the effects of the dispossession of Casiavera's land. In the 1989 image, the degraded, barren soils of the cattle ranch are evident. A second image, this one created in 2015, nearly twenty years after Casiavera's reclaiming began, shows agroforests growing precisely where the land was most degraded on the Dona ranch and plantation. Reclaimers grew agroforests across more than half of the areas that were previously barren soil.

At the landscape scale, forest cover across the collective land tended toward regrowth. Reading satellite images from 2001 to 2015 reveals the dynamism and flux of the smallholder landscape. By 2008 the upper reaches of the collective land were largely under agroforest. Susilo and his family's mixed-cinnamon forest was the largest single forested plot, covering roughly one hectare. Around this time, Wangga began to plant her plot; her cacao trees are visible, no older than three years. At the lowest boundary of the collective land, another few families' agroforests have closed canopies, indicating that these were some of the first reclaimed plots.

Then in 2010 and 2015 changes within individual plots emerged. Agroforest cultivation, harvesting of individual timber stands, clearing of unused plots to prepare them for planting new agroforests, and plots of banana monoculture stands can be seen. Several vegetable plots were planted, burned, and planted again; the timber species planted in stands on the edges of these plots continued to grow.

All these changes are ecological disturbance. Assemblages of life were altered, new cultivars introduced and others removed, all at a fantastic rate. How to explain this dynamic emergence as change that is far more complex

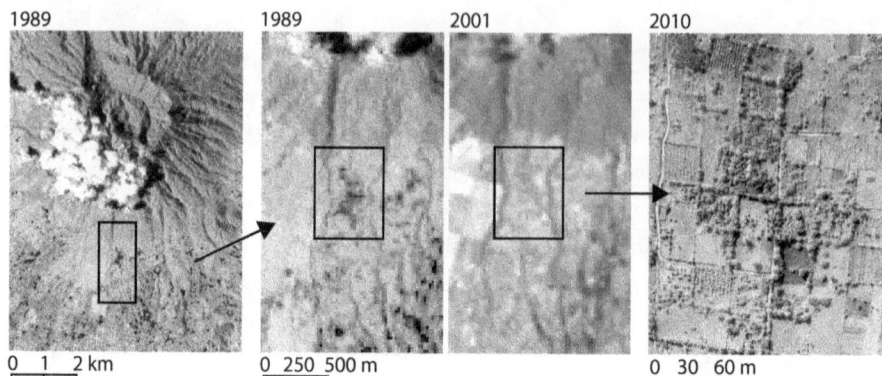

1989   1989   2001   2010

0 1 2 km   0 250 500 m   0 30 60 m

FIGURE 13. Satellite images showing forest regrowth after reclaiming the land on the Aren volcano. In 1989 the degraded, barren soils of the cattle ranch and plantation, rendered in dark grays, are visible. By 2001 growing agricultural forests, rendered in light grays, have replaced the exposed soils as Casiavera's smallholders reclaimed the land. Taking a closer, landscape-scale view from 2010 shows reclaimers' emergent forest plots. Left to right: Landsat 7 and 5 composite (bands 7, 5, and 3) 30 m resolution time series and Worldview panchromatic 0.46 m resolution image.

than simple trajectories of deforestation and reforestation? Without a socionatural theory of landscape change, any analysis risks misplacing smallholders, as they have been far too many times throughout history. Indeed, not long ago the kind of agroecological management of the land seen in Casiavera was considered to cause the destruction of ecological diversity, not be a source of it.

In a reversal, Western ecologists have only recently come to realize that temporal and spatial changes are required for biological diversity to exist. The shift in mainstream academic ecology began in 1978, when Joseph Connell drew on observations of forest changes in Uganda and Nigeria to argue what at the time was a highly controversial idea: assemblages of life are at their most diverse in areas of disturbance, not areas under any static equilibrium. When disturbances are absent, over time forests tend to form less diverse ecologies, with only a limited number of species dominant. But when under disturbance, landscapes tend toward more diverse mixed stands. Similar community ecology studies in Southeast Asia have since shown the trend to be enduring.[6]

Connell developed this "intermediate disturbance hypothesis" in mixed forests alive with smallholders, who provided disturbance through cultivation

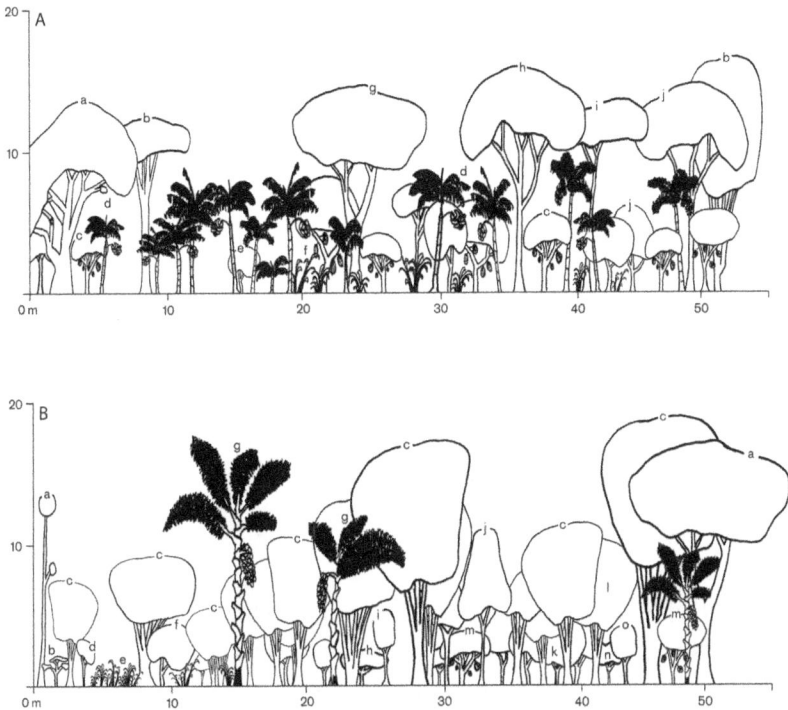

FIGURE 14. Drawings showing mixed clove and cinnamon agroforests on the collective land. (A) Species: a—candlenut (*kemiri*), b—clove (*cengkih*), c—cacao (*coklat*), d—banana (*pisang*), e—avocado (*alpukat*), f—elephant grass (*rumput gajah*), g—red cedar (*surian*), h—mahogany (*mahoni*), i—bitter bean (*petai*), j—cinnamon (*kulit manis*); (B) Species: a—red cedar (*surian*), b—trefle gos (*sikaduduak*), c—cinnamon (*kulit manis*), d—avocado (*alpukat*), e—durva grass (*rumput mangala*), f—Bridelia sp. (*kenidai*), g—sugar palm (*aren*), h—bandicoot berry (*memali*), i—rose apple (*jambu air*), j—mahogany (*mahoni*), k—dogfruit (*jengkol*), l—pink-lime berry (*sicerek*), m—cacao (*coklat*), n—Java almond (*kenari*), o—mangosteen (*manggis*).

to support the diversity of rainforest landscapes and were not inherently destructive of it.[7] Disturbance theory provides an ecological explanation for the landscape-scale diversity of Casiavera's agroforest emergence. Reclaimers' selective clearing and planting were the disturbances that undid the spatial and temporal homogeneity of a degraded grassland. With compacted and nutrient-deficient soils and a destroyed seed bank, the forests on Casiavera's collective land were not capable of regenerating alone. Like nature's windfalls, floods, droughts, and fires, reclaimers created disturbances that created new ecological processes and assemblages.

TABLE 3   LAND USE ON THE COLLECTIVE LAND PLOTS

Primary Use	Total Plots	Dominant Cultivar (n)	Plots with Livestock
Agroforest	103	Banana (44)	12
		Mixed (33)	8
		Clove (10)	1
		Avocado (7)	3
		Rubber (5)	2
		Sugar palm (2)	1
		Cacao (1)	1
		Cinnamon (1)	1
Grass	31	Elephant grass (31)	21
Unused	25	—	—
Clove	10	Monoculture	6
Tobacco	7	Monoculture	0
Vegetable garden	6	Casava (6)	0

Source: Author's land use observations on the collective land, May 2016.

The result was that a flourishing mosaic of agroforests with a high diversity of economically useful trees took root over time (see table 3). By 2016 Casiavera's reclaimers had planted a bit more than a third of the collective land as agroforests. One-fifth of plots were grass, to feed corralled livestock. Only one-tenth of plots were planted in small-scale monocultures, demonstrating an aversion to planting any single commodity crop.

With their spatial and temporal heterogeneity, these mixed agroforests mimic rainforests. Plots of cassava and banana, staple tropical food crops around the world, marked new efforts at reclaiming. Dense groves of durian, candlenut, cacao, clove, cinnamon, avocado, sugar palm, bitter bean, stink bean, jackfruit, passion fruit, and soap nut grew on the older plots. These food crops were interspersed with stands and rows of timber crops—mahogany, surian, and meranti—all valuable woods traded worldwide.

These deeply human rainforests grew into the older, non-anthropogenic cloud forests and patchworks of riverine forests, called the wild forest (*rimbo*). Mango, mangosteen, and fig grew in these older forests—traces of human influence in these forests too. Weedier species were cultivated and collected, like *melaku* (Alangium kurzii) timber. Bamboo groves contained many species of the fast-growing and easily transplantable grass, especially *bambu srik* and *bambu kapal* (*Gigantochloa sp.*). Various species of grass were planted for cattle feed.

Even as this diversity supported much more than human life, it gave people things they needed: hunting grounds, firewood, building materials, medicines, fruit, and commodities for cash. The forests gave people tools for hunting, fishing, cooking, and fertilizing the soil. They planted terraced paddies down below the collective land in which to harvest rice as well as raise ducks and collect eels. In the mornings they netted fish and collected watercress out of their dooryard ponds. After rain they gathered oyster mushrooms in the darkest corners of their agroforests.

When the volcano wasn't shrouded in fog and mist, they searched the highest reaches of the cloud forests for game—mostly birds, mountain goats, and deer—as well as honey, mushrooms, and medicinal plants.[8] Joining the human work were the morning calls of the Sumatran lar gibbon and the resounding knocks of the great Indian hornbill to give the agroforests a wild character.[9] But this was no "relic" of tropical rainforest. Over time, as reclaimers cultivated tree crops, a diversity of life beyond the human thrived. Humans interacted with plant cultivars and thousands more species of life in an ecological territorialization where reclaimers consolidated their land claims and solidified their smallholder economy.

Successful plots like Wangga's integrated multiple living species under human manipulation through space and time. The Casiavera landscape became an agro-silvo-pastoral assemblage, meaning people planted annuals like vegetables and grasses, cultivated tree crops, and raised cattle all at once across their small plots of land. Reclaimers worked as a family unit on small plots of land, producing not primarily for subsistence but for profit. When they were not farming, they were working construction, trading, or working on others' farms for wages. Diversity and complexity allowed smallholders to adapt under the rapidly shifting conditions of weather, world markets, corporate-led commodity speculation, and state planning.

Such small-scale, flexible ways of working the land, observed across the world, are especially capable of reducing agrochemical and energy inputs, including human labor, to a minimum. They are well suited to marginal soils and climates. They also tend toward horizontal distributions of production (because of their small-scale and lesser reliance on export commodity crop market hierarchies). Within Casiavera's agroecological mosaic was a patchwork of plots where different levels of intensification occurred.

While intensification in the Bukit Barisan typically involved deforestation and high-input, large-scale monocultures, on the volcano a different,

agroecological intensification of another kind unfolded. When fully realized, agroecology sustains some of the most productive farms known, with
both total yields per hectare and time invested surpassing those of industrial
monocultures. While industrial methods have driven increases in the gross
quantity of agricultural production, there remains no evidence that total
productivity, measured in yield per area, of large-scale plantation agricultures are higher than smallholder cultivation.[10]

In large part, agroecology's productivity is due to a kind of intensification that minimizes human labor and maximizes agricultural use of space
above and below the ground. A four-tiered cropping system makes use of
tree crops from 150 feet above the ground all the way down to the understory
and even a few feet below the ground, with root crops and fungi. Conversely,
monocultures are limited to one tier in space, be it the canopy level for oil
palms or on the ground for soybeans.

It is not surprising that Casiavera's reclaimers took up agroecology, given
the host of challenges they faced. The state still labeled their occupation of
plantation land de facto illegal. Without state land certificates, reclaimers
didn't have access to most of the government agricultural programs that
subsidize chemical-input-intensive, smallholder monoculture cropping techniques. Nor could they receive state irrigation funding to develop rice paddies. They had little access to credit. Their land access was limited. It was
by necessity that the reclaimers set out to make the land productive without loans or purchasing agrichemicals, commercial seed stocks, or machinery. Instead, reclaimers turned to principles of diversity, multiple use, and
energy cycling and embraced the ecological processes of a whole host of
other living organisms.

Reclaimer's agroecology began with what they did have on hand: seeds.
Lufti was one of many reclaimers who went up to the cloud forests to collect wild cinnamon and surian seeds for planting. People also gathered them
from the home gardens down below. In this way reclaimers avoided paying hundreds of dollars at nurseries. Lufti germinated tree seeds along a
shaded wall of his house and transplanted them one by one in his plot as
they matured. The task took many years but did not require much capital
or heavy labor.

There was a continuous movement of seeds within the community. They
were most often shared freely, not sold. Farmers who noticed a particularly
nice-looking tree while passing through a plot harvested a few seeds without

FIGURE 15. Lufti with his tree seedlings.

asking. Lufti told me he collected durian seeds from a tree nearby that was not too tall but had the biggest, sweetest fruit he could find to plant in his plot. One day word spread that Lufti's wife, Wera, had a few pineapple succulent starters, and I joined Den, a longtime Casiavera resident originally from Java, on an hourlong walk to Wera's home to get a few of them so we could plant them in Den's home garden.

In the coffee huts and cooperative meetings, reclaimers discussed their seeds' successes and failures to identify the most locally adapted ones. Plant genetic material joined knowledge to move across and through a network of friends, neighbors, farmers' groups, and government agricultural extensionists. This stream of people-plant relationships existed parallel to and free from agribusiness' strictly controlled industrial plant knowledge and seed clones.

Missing across Casiavera's agroecological landscape were other components of agribusiness as well, most noticeably machines. There was little mechanization on the volcano beyond the automobile. From an agroecological perspective, machines require capital and energy, often break, only have one use, and so are wasteful. Take for example the rototiller, a hand-operated, gas-powered tilling device shaped like a wheelbarrow used to turn over rice paddies after harvest. As an industrial replacement to the water buffalo and plow, the rototiller is a piece of machinery that is often found in rice-cultivating communities across Sumatra but is not popular in Casiavera.

When Lufti's father ploughed the rice paddy across from my home, I watched as he spent two complete, very hot days in the mud to do the job with a water buffalo under yolk. Finally done, dark brown from the sun, and covered in grey mud and sweat, he sat next to me on my covered patio with the relaxed smile of a job completed. I asked him why he did not use the rototiller instead of his plow attached to his water buffalo with plastic rope and wood harness. He shook his head like all older folks do at the silliness of the younger generation:

> That machine, I bought one when they first came out. It breaks. And it does not go as deep [into the soil] as my plow. In just a week or two the weeds will grow back. The gas. The repairs. It just eats up your wage. And the machine is really hard on your body, it uses too much of your labor. It is not easier. My buffalo, she will come to me if I call her, just sitting here, but you must transport the rototiller.

As Lufti told me, the water buffalo is cheaper than the machine, free if you can breed your own instead of purchasing one. Unlike many of the machines available for purchase to reclaimers, water buffalo are known for their hardiness and fertility. Although they are not carbon neutral, they are powered on widely available grass and the stubble of cut paddy instead of fossil fuels. And unlike a machine, they can be eaten when they reach the end of their useful lifespan if one is so inclined.[11]

Smallholders have often been singled out as luddites. Part of the reason workers often do not like machines is that machinery is typically designed to displace human and other animals' labor, not support it, as in the case of the rototiller as well as the combine harvester, a machine long held to be the smallholder's antagonist from the US Midwest to Southeast Asia. At other times workers reject or attack machines because they are taken to represent capitalism itself, as in the case of workers sabotaging plantation equipment.[12]

When new technology is only a means to a capitalist end of accumulating profit, not an improvement of workers' livelihoods, avoiding machinery makes complete sense. This is often the case with the equipment agribusiness manufactures. In the United States, as mechanization increased before and after World War II, agriculturalists' economic position was eroded. Smallholders became enmeshed in onerous debt relations to the purveyors of agricultural machines. The single most important reason that the purchase of farm machines created ruinous debt for farmers is that there is no evidence that mechanization increases intensification, measured by yields, and only at times does it lower the cost of producing food as compared to employing human labor.[13] Thus, the agribusinesses' promise that farmers' payment for machines would increase yields and profits most often went unfulfilled.

Smallholders' aversion to machines, agrichemicals, and corporate seed stocks has led far too many agronomists, farmers, and politicians to conclude that agroecology is archaic. It is true that Casiavera's reclaimers' agroecology is an extension of practices of forest manipulation and tree cultivation that took root in Sumatra some fifty thousand years ago.[14] Yet Casiavera's agroecological landscape came into being in response to the difficult lessons of over 150 years of agrarian capitalism, so it is modern, not pre-capitalist. It is also not receding under the forces of modernization, but rather still emerging and consolidating. A political struggle brought new relations to the land. Then, a reconstructed agroecological praxis made it productive. It is a landscape defined by human actions, an agroecology full of markers of social life and nature. The changes across the land were not unilinear. There were multiple losses and gains of forest cover over time.

The Dutch imposition of *erfpacht* corporate control in 1914 destroyed Casiavera's peasant forests. Then Casiavera's smallholders, caught up in the earlier revolutionary-era reclaiming movement, began creating the forests anew. Satellite photography shows how the forests expanded in a patchwork up until 1968, never becoming a closed canopy like the cloud forest higher up on the volcano but growing nonetheless. Then, the New Order ended smallholder control and the Dona Company cleared the land for pasture and ginger monoculture. Three decades later, reclaiming the land involved its reforestation. In some areas on the collective land the trees have grown to create a canopy that spans plots, bringing the forest a connectivity it has not had since before colonial rule.

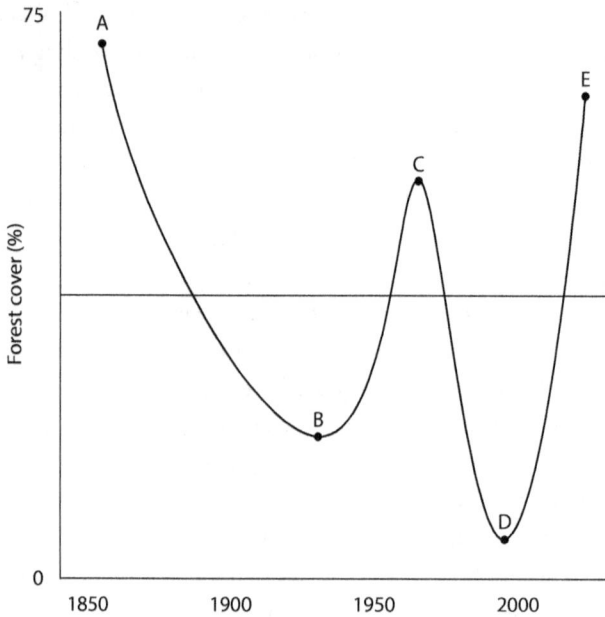

FIGURE 16. Graph depicting trajectories of land control and landscape change in Casiavera. A–B: forced Dutch coffee cultivation and *erfpacht* corporate control of the land destroys the peasant forests, 1875–1940; B–C: smallholder use of the land replaces colonial control during the Indonesian Republic (1950–67); C–D: the authoritarian New Order ends smallholder control and turns over the land to the Dona Company in 1968, after which the land is converted into pasture and tobacco and ginger monoculture until 1996; D–E: Casiavera's reclaimers gain control of the land to cultivate diverse agroforests.

As remarkable as the ecological function and diversity of reclaimers' forests are, these cultivated forests take meanings beyond ecological life cycles. They are the form that women and men like Wangga and Lufti created to counter their own dispossession. Wangga and the others expressed their power as part of the Minangkabau matriarchate across the landscape as agroforestry livelihoods of farming and trading. Building on the knowledge of agroforestry contained within their own matriarchate of tending trees like cinnamon, clove, and coffee as commodity crops, Wangga and the others reconstituted their relations to the land and contributed to an emergent smallholder economy, upholding the emancipation of Casiavera's reclaiming movement.

A MIND TOWARD POLYCULTURE

During the long decades that Casiavera's Minangkabau peoples survived the W. H. Samuel logging and cattle operation that began in 1911, then again under the Dona ranch and plantation that began operating in 1968, a combination of state and corporate command radically altered ways of social life. These rulers inscribed a "monoculture of the mind," an erasure of smallholders' existing *metis*, their collection of cultural (and agroecological) knowledge, in order to impose discipline and industrial forms of monoculture planting and labor.[15] Much knowledge of agriculture, ecology, and nature was lost. Hence part of Casiavera's reclaiming required reconstructing agroecological knowledge, an effort that came to privilege diverse agricultural plots, or polycultures, above the plantation's paucity of life and ongoing, slow violence.

Rudi, a pot-bellied and deeply tanned forest farmer with a plot in the collective land, saw his agroecology practices as taking a place within the origins of his culture. According to Rudi there is nothing foreign about agroecology; it is very old and endemic to the Bukit Barisan. "Our philosophy has always been in the buds of agriculture (*pucuk tanaman*)." Rudi explained, "We take our inspiration, our knowledge, from the forest. So, we base our agriculture on what we see in the forest. Just the same way that however many years ago the peasant emerged from the forest to do agriculture." He repeated an old refrain, "Growth in nature is our teacher" (*Alam tarkambang jadi guru*), to describe the formation of his agroecological knowledge. He "reads nature like a book," as he said. The natural world is both substrate and blueprint for Casiavera's agroecology.

While not completely erased, this knowledge was fragmented and diminished through Casiavera's reclaimers' history. "You could say it was capitalism," Rudi told me, that disrupted their peoples' agroecological lifeways. Rudi schooled himself in Marxist structural analysis as part of reconstituting this agroecological knowledge. "Our human resources are still low here," Rudi lamented. The land and cultivars required for planting were present, but the knowledge and social relations of production required for them to flourish were only recently taking form. This time, smallholders' tendency toward polyculture emerged out of their long, troubled relationship with colonialism and industrial monocultures. Theirs is an anti-plantation agriculture, founded in opposition to monoculture.

Casiavera's reclaimers brought their memories of commodity crop busts to the creation of this anti-plantation agriculture. Laboring under Dutch overlords in the coffee rows of forced cultivation was not something quickly forgotten in Casiavera. The people who experienced the old, colonial way of life had passed away. But the memory of forced cultivation remained. It is this memory of coffee's harms as a monoculture commodity crop that underpinned Casiavera's reclaimers' aversion to cultivating coffee in particular and monoculture in general.

For Yono, a sixty-six-year-old grandfather, sugar palm tapper, dragon fruit cultivator, and shaman, coffee was "too much work," even though coffee can easily be integrated into agroforests with little work. Like most everyone else in Casiavera, he did not cultivate the crop. Similarly, for Lufti—a young man who planted thousands of trees on his wife's land; tended many stands of clove, rubber, cinnamon, and chocolate; and worked dayshifts as a cow herder and the odd construction job—coffee was "too difficult." Views like these are the memory, the cultural afterlife, of the ecological, economic, and cultural suffering brought about by early capitalists' fetishization of coffee as a commodity crop.

One evening, overhearing me wondering why no one cultivated coffee on the volcano, a woodsman, Midwal, told me that his grandparents were the last to cultivate the crop. "How did they live?" I asked him. He had not experienced it, Midwal responded, and so he could not answer my question. His parents had only mentioned to him that his grandparents were forced to cultivate coffee on the volcano, and his parents grew it above their homes. It is not something that his family dwelled on. But Midwal offered to show me where their coffee still grew.

A week later Midwal took me up into the forest.[16] We made our way up the volcano's slope on one of the many trails that woodsmen from Casiavera maintained throughout the forest. A short rise at the start of the forest flattened out onto a wide slope. Midwal, still lean but with a pronounced limp at his advanced age, pointed out a young coffee tree in the undergrowth of the canopy, shaded by a massive wild guava (*kalik jambu*) tree above. "This is Robusta!" He gestured for me to look around, "The coffee is like a flood in the forest here." Across the understory coffee trees of all ages were growing. Some newly sprouting saplings were just emerging out of the volcanic soil. Others reached fifteen feet tall. "It came from planting, during colonialism (*penjajahan*). Now it is scattered all over by birds. The coffee did not disappear. It is in the forest!" Midwal and the other reclaimers didn't tend these coffee trees growing in the cloud-forest, because the New Order made this part of the land a protected forest (*hutan lindung*) where hunting, logging, and agriculture are illegal. Still, Midwal was often in the forest to hunt or cut timber, and when passing through he would gather a few handfuls of ripe coffee fruits. Coffee's traces among the landscape assemblage remained, just as markers of forced cultivation remained in Casiavera's smallholder praxis.

Social relations of production allow certain crops, and the agricultural systems they are produced in, to become sources of exploitation, emancipation, or something in between. In the case of the forced coffee program, planting coffee trees brought a loss of sovereignty, land control, and autonomy. Casiavera's smallholders never returned to cultivate coffee. The collective memory of the failings of the monoculture model, this afterlife of coffee as commodity crop, remained to give Casiavera's twentieth-century reclaimers the greatest benefit: Casiavera's reclaimers' preference for diverse, ecologically attuned polycultures.

As my time in Casiavera flowed from days into months, I was struck by just how Casiavera's reclaimers elaborated the value of diversified production. Crops needed to "fit together." The best examples of agroforestry were "complete" (*lengkap*). This was a positive assessment of the kind farmers like my own grandfather would have called "productive" or "profitable."

Diversification was said to sustain economic and energetic cycles: "If the rice fails, we have our cassava, our corn, and our cows to sell," reclaimers often told me. Their banana stalks could be used to feed their cows, the cacao pods were fed to their water buffalo, and their peanuts grew up on cornstalks as climbing trellises. These are the energetic cycles and lifeways that create diversification. Without organic fertilizer made from smallholders' cattle manure, tree crops would not have taken hold in the degraded soils. Without the cultivated agroforests, extractive pressures on the cloud forests would be so intense that the forests would not exist to supply the water the cows and commodity crop trees need.

Reclaimers diversified their smallholdings even further, working to create access to rice terraces, either by kinship, private ownership, leasehold, or sharecropping, to complement their agroforests. For the more than half of reclaimers with access to rice terraces, their agroforests were less a source of self-provisioning and more a source of income. These tree crop incomes, in turn, reduced families' risks from a failed rice harvest. A decentered assemblage provided smallholders greater flexibility and more options, the kind of freedom that underpins the concept of autonomy.

The Sumatran reclaimer, peasant union leader, and agrarian scholar Julius Julian Polong has spent more than three decades discussing and refining agroecological praxis in the Bukit Barisan, including at his own smallholding squat. Polong has written that agroecology is nothing less than the harmonious integration of humans and nature across the landscape. For Polong, the "edge effect" must be respected, because it is at the boundary of two different ecological patches where the greatest "energy" and "diversity of life" are contained. To contain as many edge effects as possible, polycultures must be "human scale." Such a scale also allows the manipulation of the "relative location" of all tools, non-living ecological processes, and life forms to maximize mutual benefits. All praxis elements should support "natural methods and processes" with "multiple elements and functions." Processes of "natural succession," noted with observation and research, inform future agroecological practice. Such an approach minimizes external needs and emphasizes overall stability through plurality. "Energy planning" and "energy cycling" should minimize energy demands, especially human labor, and "reuse and recycle all resources as many times as possible."[17]

Polong's view is that agroecologists do not seek to transform the world to overcome or conquer nature, but instead seek to reveal the ecological processes and functions of the landscape. By doing so they seek to harness forms of ecological energy and social power. Diverse human-plant interactions cement a perspective of nature in which human action is just one stream of action, energy, and influence among a much greater collection of life.

Casiavera's reclaimers existed in a commonsense world where their agricultural skills, or to put it another way their human-plant relations, informed their livelihoods and environmental perceptions. To cultivate their agroforests, reclaimers focused on the plants, pollinators, and predators around them and the way they could alter the abundance and spatial distribution of these life forms.

A walk through the fields was time well spent, even if little "work" was done. Exploring and moving through the land can be an experience akin to spending time with relatives. Attending to the landscape in this way creates a physiological connection to place and the ancestors that moved through it before. This sensory experience takes place within a single, unified entity (a *landscape*), constructed out of an accumulation of knowledge of plants, animals, rivers, ridges, and foot tracks. It is as intimate as any experience in day-to-day life, with its own place alongside sex and other multisensorial social encounters as the most richly emotive human experiences.

With a finely attuned perceptual knowledge, skill can emerge; the greater the skill, the deeper the richness and creativity with which nature can be attended to.[18] In creating their own smallholder economy under difficult conditions, reclaimers did more than rely on the strategic deployment of programmed responses to external environmental factors. They relied on planning, care, and nurturing. All required relationships with other living beings, not limited to the human. These are skilled and attentive behaviors designed to cope in the world.

An agroecological multisensorial knowledge, or aesthetic, can be described as a form of creativity that builds on a repertoire of botanical, climatic, and economic understanding and experience. There is trade, movement, and transformation. Deep enskillments are gained with experimentation and learning from nature and others.

Agroecology immersed people in assemblages of life richer than most. Through this immersion Casiavera's reclaimers laid claim to a modern, agrarian subjectivity. A new kind of agroecological knowledge was acquired, superseding previous knowledge of coolie laboring. A long-circulating way of thinking about the land became tinged by concepts of anti-capitalism and anti-monoculture, part of the *metis* accumulated through daily interactions and experimentation. This new agrarianism contains a heightened awareness of not only rural work, but also the fate of the urban underclass. When discussing the full range of livelihoods people lived in their time away from Casiavera, reclaimers often reminded me that the urban poor live under bridges, in shanties up on stilts above toxic urban river drainages, and even in Jakarta's vast trash yards. Access to land not only provided economic productivity; it offered reclaimers a nice place to live.

Avoiding single-purpose machines, not tilling the soil, refusing to purchase toxic inputs like pesticides, opting out of cycles of debt, and growing a diversity of fruit crops not dominated by agro-industry are not apolitical economic decisions. Casiavera's forests carry a moral component, in that people in Casiavera are aware that what they are doing is rejecting an "investor economy" and is considered by outsiders to be "good for nature." The kinds of social action Casiavera's reclaimers are engaged in are motivated by green concerns—pollution, soil degradation, and other ecological grievances—but their agroecological politics attests to how environmentalism has a panoply of positions and uses beyond those most Westerners conjure up, including concepts of work and labor. On one level Casiavera's agrarian discourse fits with existing transnational environmental movements' strategic representations. On a second, more fundamental level, Casiavera's ideas of nature and work have restructured ecologies to better support their own visions of autonomy and livelihood.

A polyculture of the mind underpinned the diversification and flux of the agroecological landscape. Plants and people changed each other, for mutual benefit. For this reason, Casiavera's reclaimers often placed their agroecology within its true context—nature—much more deeply than any materialist, utilitarian way. They instead took the bold position that humans and diverse forms of non-human life can coexist without exploitation.

These ideas exist as part of an agrarian cosmology that circulates through Indonesian society. A strong thread in these ideas is that alienated, non-harmonic human behaviors lead to the destruction of both humanity and ecology. It is not surprising that in Aren, long a site of Minangkabau matrilineal society, many speak of nature in terms of fertility. As the matriarchs (*ninik mamak*) of the family lineages remained in control of many matters of land and agriculture to nurture their families, the natural world was part of women's agricultural, economic, and familial roles. Less expected is a perspective that emphasizes the human position in nature to be one of continuous, cyclical creation and destruction. As one reclaimer, Menhir, told me, "Nature was first! Nature was here before us, and will be there after us."

Menhir's ideas of existence as connections between humanity, nature, and the creator that are unstable through time made me think of how Sumatra was born and remains engulfed in volcanic fire. Periods of calm are punctuated by lavas that destroy and then cool to form the hundreds of volcanic peaks of the Bukit Barisan range. Rising north-south along the length of the Sumatra, these same peaks create altitude-charged thunderstorms that form the watersheds that sustain the tall forests and wide rivers of Sumatra's east coast—the Kampar and Indragiri that cut across the eastern edge of the plain to tumble down the Bukit Barisan's deep canyons, in rushing rapids and falls, carrying life-giving volcanic soils out into the 150-mile-wide, swampy eastern coast that eventually became Sumatra's colonial industrial plantation belt.

The constant, volcanic remaking is refracted in the social conception of creator as ambivalent nature, with phases of fertility and destruction. Nature can sustain life but can also destroy it with natural disaster and social ruin if human mismanagement pushes the landscape matrix out of balance.[19] As one forest farmer on the collective land told me, industrial agriculture may be part of modernity, but it hinders nature (*halangi alamnya*). It was because of this imbalance that the reclaimers were unwavering (*teguh*) and assertive (*tegas*) in their insistence that the land be used for smallholder agroforestry.[20]

Images and stories of such agrarian cosmologies travel with the farmers, educators, activists, and artists who produce them. Visual arts are especially popular for their ease of interpretation and circulation. Within Indonesia's agrarian justice movements, the Javanese artist Surya Wirawan is known for using woodcuts to illustrate these agrarian cosmologies. His experiences led him to help start an activist-artist collective, Rice's Fang (Taring Padi), which became an active crew in the new social movements in the early 2000s.

Surya's work gives a view into an agrarian well-being within and among diverse natures. In *Humans Preserve, Nature Provides* (*Manusia Memelihara, Alam Memberi*), the artist depicts a young woman holding an overflowing basket of fruit and vegetables, standing in front of a dirt road that travels through fields and fruit and rubber forests.

Sitting down with Surya in his studio in Yogyakarta, Java, he told me that as an art student under Suharto he was so poor that he lived in the basement of a building on his university campus alongside scores more students, who spent years studying and sleeping there on tabletops to stay off the dank cement floor. His grandparents were all farmers living in the countryside, so from an early age Surya was also aware of the agrarian suffering of the late New Order.

Humans Preserve, Nature Provides came out of a workshop Surya participated in with Taring Padi at their cooperative art studio, gardens, and guesthouse. They convened in celebration of Earth Day to spend a few days working side by side. Surya planned to create a different work, called *Humans Crush Nature, Nature Will Punish People*. Yet after the first day of the workshop, Surya felt there was already too much suffering, with persecution and struggle everywhere and so many campaigns lost. So Surya made something more positive. The message he chose carries the same meaning as the negative campaign slogan he rejected; the inverse of the idea that

FIGURE 17. Surya Wirawan's 2004 woodcut print *Humans Preserve, Nature Provides (Manusia Memelihara, Alam Memberi)*.

if humans preserve nature, it will sustain them is that if humans destroy nature, it will destroy them.

Found at each end of the telling of this message, according to Surya, is the idea that "nature and humans must have a balance. Without this balance everything will be stillborn (*keguguran*)." In a universe where the world is born and dies, again and again, in an infinite cycle of creation and destruction, the loss of "balance" can destroy the life cycles of the universe, leaving only death and tragic non-reproduction, a world "stillborn" and unable to grow:

With [rural] industry today we do not have any balance. The point is that we need what industry brings us, but there is no balance. Palm oil, for example, is good, but we need a balance. It is the greed that brings us out of balance, their greed. Agriculture is good, but industrial? I don't know. I don't see any balance with what they are doing in Sumatra, Kalimantan, in Java. . . . Food is provided by nature, we have trees that give us life, the rain, the rice fields. We need the balance, where none is higher than the other. The animals, the plants, the humans, the trees, we are on the same level.

Running counter to the kinds of desire that allow for a holistic agrarian well-being, Surya sees greed and industrial plantations. For in Surya's concept of balance is an idea of equality. He sees the only way forward, beyond dispossession and capitalist extraction, as involving restoring balance by valuing equally all parts of the countryside.[21]

Surya joins Casiavera's reclaimers to contribute ideas and work that give form to modern smallholder landscapes. Located at the fulcrum of reclaimers' survival is their agroecology. Indeed, there is nothing less than an agroecological revolution—epistemological, technical and social—afoot.[22]

At its broadest, agroecology is a bricolage of diversified work that mitigates the multiple risks smallholders face in global markets. Instead of entering into onerous, dangerous, and not-very-profitable jobs that are fixed in routine and place at the plantations and factories, smallholders often choose a mobile life that balances agroecology, non-agricultural entrepreneurship, and waged labor. Rather than approaching the farm as a factory, with the worker required to constantly attend to an assembly line of monocultures, agroecological ways of thinking and acting allow rural workers the freedom to leave their fields for extended periods. This freedom from the field allows smallholders to also work as laborers, office workers, and traders, or like many movement members, as advocates, writers, and artists, all while their agroforests grow.

Modern smallholders are more than likely than not to have a hodgepodge of work, varied across the course of their lives as well as the seasons. Workers combine rural and urban, informal, seasonal, and waged labor with cultivating agricultural crops. Smallholders do urban work, like trading, management, and retail, alongside agricultural labor. They also do rural work that is not agriculture, like construction, wildcat mining and logging, driving, and auto repair. This diversification is not necessarily a sign of individual smallholders' transition away from agriculture; this bricolage is part of the effort to retain or create anew a smallholder economy. Smallholding is making use of whatever comes to hand.[23] It is constructing work out of what is available. Bricolage is a trajectory of ecosocial change that smallholders see as an autonomous solution to the limits and ills of joining the industrialized proletariat.[24]

Certainly deprivation is one reason some smallholders have come to be only "it depends" farmers, as Indonesians call the bricolage rural worker. On plots too small, ruined, and remote, there is no doubt that smallholders can be forced out of a life of full-time agriculture by their inability to turn a profit or even sustain themselves and their families from the land alone. For the most marginalized, part-time farming has often led to depeasantization and proletarianization. It is usually already better-off smallholders who improve their own lot with livelihood diversification. Most perniciously, as plantation and factory economies consolidate, their owners and managers often seek to access rural labor more cheaply by co-opting smallholders' own food production as a subsidy to their corporate below-living wages.

Still, livelihood bricolage is more than a survival strategy of last resort for the underclass. Diversified collections of work centered around the family farm sustain rewarding smallholder lives. Take Lufti's experiences. A tall and lean father of two, Lufti was one of Casiavera's most energetic agroecologists. He planted hundreds of commodity crop trees on his wife's and mother's lands and hundreds more for a wage on others' plots. As we walked through his family lineage's lands below Casiavera, Lufti told me that he first learned to cultivate agroforests from his parents, working with them in their home gardens on their time off from laboring on the Dona ranch. The knowledge came from a long time ago, from their "ancestors." Lufti's agroecological knowledge deepened when he came to work with the local branch of the Indonesian Peasant Union. Lufti took up the peasant union teachings to see tree planting as a way into the smallholder middle class that requires relatively little capital or labor. Along with this appreciation of small-scale, diversified agroforestry came skepticism about debt, landlordism, and monocultures.

It began to rain. We waited it out in a storage shed on the family plot. Over the roar of the downpour hitting the zinc roof, Lufti elaborated for me the importance of the bricoleur approach: "After planting mahogany I can be a 'it depends farmer' (*petani tergantung*). I can come and go. I just come [to my plot] every once in a while. I can harvest my cinnamon bark when the price is right. Or a mahogany tree when I need cash." While his trees grow and increase in value, Lufti can leave for the city, labor on a construction site, or spend his days tapping palm sugar. Lufti tells me, "These trees are money in the bank. They are for my future, for my daughter's university. What else am I going to do when I am sick. Or when I want to retire? I will sell my trees." By maintaining a complex portfolio of crops that came to harvest on monthly, yearly, and decadal timescales, along with wage labor, Lufti was able to work toward well-being across the varying timescales of his and his family's needs. His spice, fruit, and timber stands allowed a flexibility unmatched in other kinds of agriculture.

Reclaimers turned to bricolage to be free to lead changing human lives. They exist as a counterpoint to waged laborers in the logging, plantation, and mining concessions, who repeat unending and unchanging work tasks with all of their possible energy, as if they were reducible to living machines in the plantations' biological assembly lines.

CHAPTER 7

The Predatory Work That Remains

Life is like rowing a canoe upriver. If you stop, you will go backward
and never get ahead.

—Bajo, a landless truck driver, construction worker, rock breaker,
and new father, Casiavera, January 2016

Agoez worked for ten years at Dona, mostly tending cows. "You want to
know what it is like here? First, we go scratching in the dirt (*mencakar*), then
we eat. Like a chicken,'" Agoez told me wryly. "We do not have any experience [with agroecology]. I worked all my life for the company. We are just
little peasants." Next to Agoez's home on the collective land was the first
tree that took hold in his family's plot in 1998, when he moved to the land.
It was an avocado. He told me it was exactly as old as his already rotted
hut. The tree was still short and only produced a handful of fruit every year.

Looking at this tree next to Agoez's home, I was struck by his humil-
ity and the slow, incremental improvements that this family had made to
their lives over the previous two decades, even with the radical changes in
the political and ecological assemblages around them. Agoez and his family
were "enhancing the land" (*memutukan tanah*), as Agoez called it, crafting a
new life of impoverished land that was once a site of colonial forced cultivation. Then lives were given over to being a coolie on the New Order cattle
ranch. Eventually, Agoez's family's land provided them some shade and a bit
of income. It was still not densely planted or expansive, but it was growing.
They had fifteen clove trees, three avocado trees, and a handful of surian
and candle nut trees. The plot produced no more than $1,500 of income
a year. Agoez tapped sugar palms and worked short stints of agricultural

labor for a wage, but his family didn't sharecrop or rent land, and despite his self-deprecating comments about not knowing anything, he and his family picked up a good amount of agroecology.

At sunset I went out with Agoez to tap his sugar palms. ("It really adds to my economy. It takes so little time.") We climbed bamboo ladders up to the top of the palms, then down again after switching out the three-foot-long bamboo collecting jars. We walked fast. In fifteen minutes, we were done. Later that night his wife boiled the sap down into a rich brown sugar, favored by most in Indonesia over cane sugar.

The interlocking issues of work and land control remain the single great-est challenges to Casiavera's reclaiming. Even with their collective land control and transformation of the landscape, reclaimers had stumbles and all-out failures. Some agriculturalists threw down their hoes, defeated by the soil destroyed by decades of industrial use. Nearly everyone who was able to establish workable relations with the land turned to planting agro-forests even though the Casiavera Community Council had enacted a regu-lation against it.

The council worried that growing agricultural forests would create long-term individual smallholder ties to the land, in the process eroding the col-lective control of the land. In practice, individual reclaimers' connections to their plots did harden, but they did not treat their plots as private property that could be bought and sold. After nearly two decades there was no buying or selling of plots and only a few instances of consolidation of plots on the land, mostly through shadow transfers of plots between family members. Yet even these limited transfers of plots raised the chance of a slow and limited accumulation of land in the hands of powerful individuals.[1]

Even more troubling, some people had joined the difficult reclaiming struggle only to become contract farmers on their own plots and wage laborers in the plots of others. Hanging over the reclaimers and these challenges was also the threat of the return of dispossession to Casiavera. Dona Company continued to hold a land concession. The company and National Land Agency still considered the reclamation to be an illegal land occupation. These combined social forces did not completely derail agriculturalists' pursuit of life after dispossession. But they did threaten individuals' well-being and raise questions about who benefits from what kind of work.

TOBACCO EXPLOITATION

In 2011 a team from the Agricultural Ministry visited Casiavera. They gathered a few of the farmers' cooperatives together to tell them that the government was interested in supporting tobacco cultivation. Tobacco has been grown as a smallholder crop across Sumatra since the seventeenth century. Unlike commodity crops grown in mixed agroforests like cinnamon, cacao, and clove, tobacco is an annual that demands full sun. With no shade tolerance, it is best planted in monocultures. But with tobacco prices more stable than most commodity crops, smallholders long made a go of cropping the oily, big-leafed plant.[2]

A few weeks after the agricultural extensionists' visit, one of the family-lineage leaders in Casiavera, Datuk, wrote a proposal to the Ministry of Agriculture as the head of a smallholder cooperative he had established after the government's meeting. If the provincial government would provide the funds to build a tobacco drying shed; buy seeds, fertilizer, and pesticides; and pay for the waged labor required, Datuk would donate a plot of his extended family's land to plant the crop. He proposed that after this first state-funded crop his group would be able to reinvest the profits in another tobacco crop, thus sparking smallholder production. Datuk's tobacco proposal was funded. He planted his land once, paying his family and a women's farming collective to provide the labor. The crop was brought in. But Datuk never did sell that first crop. He ended up leaving it to rot in the drying shed. After that he closed the shed and never opened it again. During my fieldwork the carefully constructed shed sat unused alongside the empty plot of land, the door padlocked, its wood siding looking new.

Even before it was ready to harvest, a trader had come to buy Datuk's one and only tobacco crop. He offered Datuk a price. Datuk declined. "Too low," he told me. "That price would have only been enough to plant the next crop." I asked if he had looked for other buyers to get a better price. Datuk looked at me with a long pause, shook his head in frustration, and then only told me, "No, the entire thing was not optimal." I was a bit confused by Datuk's choice of an answer and reticence about clarifying for me. It seemed to me it was a great deal less than "optimal." He had lost an entire crop, and his project never got off the ground. I imagine the agricultural extensionists were not too happy about the failure, either. I pushed Datuk a bit more. Would he try again, to plant tobacco? "No, because I have to sell to that trader. And he takes too much of the profit."

I understood that for Datuk there was a very personal relationship underlying his feelings about this particular trader. The trader was actually what the reclaimers called a tobacco "boss" (*bos*) in the area. The man owned one of the only tobacco warehouses in the nearest city, giving him a great deal of influence in the local tobacco market. He was much more than a trader of tobacco, however; he was also a contractor who spent much of his time traveling the countryside looking for willing smallholders whom he could sign up to plant a tobacco crop that they would sell to him, so that he could fill his warehouse, where he dried, cut, and sold the crop to wholesalers and exporters. In effect, the boss was a classical monopsonist: a single buyer in a market of many commodity producers. Across the volcano, tobacco planters became price takers, forced to sell their crop at whatever price the boss offered.[3]

Datuk knew all of this before he wrote his proposal. He had even planted tobacco before, after the community started cultivating the collective land. That time Datuk took a loan of around $600 from the tobacco boss, from which the boss subtracted the cost of the required seeds, fertilizer, and pesticides that he gave to Datuk. In return for this loan Datuk agreed to only sell his crop to the boss, at a price they determined at the outset in a quintessential smallholder contract farming agreement.

Datuk says he was lucky. After this first crop allowed him to just about break even, his second harvest was a good one. He used the money to end his relationship with the boss, debt free and with a bit of money in his pocket. Even after good rain and much diligent work, on his first try with tobacco Datuk could not accumulate any cash. But he was fortunate then to have a bumper crop that allowed him to pay his debts, leave tobacco's debt cycle, and try different livelihoods. His new idea with the government project was to source the capital he needed to bring a tobacco crop to market without working with the boss. However, Datuk had failed to envision the end game; once he had a crop on hand, he found out just how difficult maneuvering within a monopsonistic market could be; he could not find any buyers other than his old boss.

Datuk was not alone in his desire to make a go of it with tobacco. More than a handful in the community had become contract farmers to plant the crop with credit from the "boss" and take losses. During my fieldwork one older couple was planting their fifth crop, still hoping tobacco would earn them a profit, or at least pay their debts. Lufti, who had planted one crop and then simply walked away from his debts to the boss, told me in his typically easy-going way:

> Actually, the whole thing is not that great. Yes, with tobacco we don't have any way out. Without the ability to choose who we sell to or what price we get. But we do it because we need the seeds, need to pay for workers. We do not have capital (*modal*).

Lufti learned about the dangers of contract farming tobacco from the peasant union meetings. These discussions teased out how to recognize that some people were in an exploitative commodity cropping relation. Even so, his friends and family around him kept going back, growing another crop, thinking that if they got lucky with a bumper crop, they would get out of the debts they had accumulated. But they didn't know the market price, and the boss was the only buyer, so they became price takers. Lufti continued:

> We do not even know how much profit that person [the boss] is making off us. If the harvest is not good, if there is not enough sun and rain, we make nothing at all. We can't eat tobacco, right or not? We can't even smoke it ourselves; it is all promised to the boss.

Echoing back to me the deep anthropological literature on debt and contract farming relationships, Lufti's and Datuk's experiences show how debt is a powerful social relation, one that predates capitalist relations, working as a tie of mutual obligation between smallholders and their patrons and other powerful people. Debt can create relationships that bring mutual benefits for both the merchant and cultivator over the long term; debt relations need not be as exploitative as Lufti and his peasant union claim.

Tobacco, especially, is a crop that is often cultivated with these long-term relationships of debt. There are biophysical reasons for this, not related to predatory monopsonists. One is that tobacco is regarded as a risky crop, sensitive especially to drought. A smallholder's debt relationship with a capital-rich trader allows her to plant again after a crop failure, rather than be left with nothing at all. But when the vagaries of nature ruin a crop, it is the farmer who is left with the debt, not the capitalist.[4]

Lufti's fully developed agroecological praxis, honed through experience and years of SPI training, enabled him to see the broader problems with the crop:

> Yes, I might invest $900 and get $1,500 back. But I may only get $800 back. And it takes so much work, and we are stuck with only one agent (*toke*). . . . And it is difficult with the chemical fertilizers, they are expensive, and not natural. They can be dangerous for our children. I will use them if I am given them from the government, but I will not buy them. But I will not use any pesticides. None! Never, they destroy the soil. They are dangerous to us.

In rejecting tobacco cropping, Lufti expressed the central tenants of agroecology: the danger of any single crop's price fluctuation and the need to reduce external inputs, financial as well as petrochemical. Capital poor, Lufti developed a deep distrust of debt. He would not take on any at all. ("I fear those payments. If we have lots of rain, or a drought and our crops die, we are stuck with those debts we cannot pay.") Lufti talked with pride about how he avoided debt by carefully using the cash he had. And he never did something as luxurious as buying a new motorbike, like his younger brother, who bought a high-powered street bike on easy-to-get but high-interest credit after a few months of working as a gas station attendant.[5]

While Lufti did not think there was anything wrong in working for a wage (he sometimes worked as a laborer on others' tobacco plots; "that is money in my pocket that day"), he did think accumulating debts was to be avoided. Purchasing pesticides could be a source of debt, but Lufti did not reject these inputs for this reason alone. He recognized their biological toxicity as well.

Another resident, a recent university graduate in animal husbandry, was more critical still of tobacco. With tobacco, the young woman told me, "We become coolies on our own land." The cultivator's position in the market becomes one of marginalized peasant being taken advantage of. The university graduate told me they became something they did not want to be, what she called "laboring peasants" (*petani buruh*), whereby smallholders work for others, even though they themselves control the means of production, their labor time and the land.

Lufti's experience with tobacco joined with time spent in the peasant union meetings to give him an even larger and more critical vision of the regional political economy: "There are no peasants in Indonesia. Only peasant laborers." Whereas Lufti and others thought of a smallholder farming family as one with the income and assets required for a measure of well-being, a laboring peasant works waged labor (*upah*) on their own land for a more day-to-day survival. Another SPI member, Mansur, who was one of Casiavera's more important traders of fruit, agreed" "I threw away a bunch of money on tobacco." Mansur thought the only reason people planted tobacco with the agent was because of the flip side of debt, access to credit. People wanted it. For most, the tobacco agent was their only source of it.[6] But according to Mansur the rates offered by the boss were not fair.

The boss played the "traders' trick," Mansur told me. The boss comes and signs up one smallholder to plant a small crop. When that crop comes to harvest the trader will buy it for a high price, and the smallholder will do well. After others nearby have heard about the first farmer's profit, the trader returns to sign them up and gain access to their land. The boss will offer a lower price to them than the boss extended to the first smallholder, and the people may not have heard what price exactly the first smallholder actually got, so they will go ahead with the planting. But they end up with debts while the agent profits.

Mansur's was an informed worker's position on rural capitalism that highlights its predatory nature. In this framework, which resonates with the outrage in Casiavera, contract farming arrangements emerge when smallholders control the means of production (land) but do not have access to the required inputs (capital) by themselves and so end up becoming dependent upon capitalists who can be exploitative. As propertied laborers, contract farming smallholders are found everywhere there are agriculturalists.[7]

Casiavera's reclaimers' feelings about tobacco revealed the dangers of debt and the way this most fundamental of economic relations is often deployed in agrarian society. Taking on debt can bring opportunities, like starting a business or owning a house. But under conditions of smallholder commodity crop production, the combination of a monopsonistic commodity chain and the extension of debt created the conditions for exploitation.

When reclaimers found themselves with debts to the boss piling up, not many followed Lufti in simply walking away from their debt obligation. Lufti called the boss's bluff because he was young, strong, and had a lot of friends. He felt that he could handle the risk that the boss would try to intimidate him or take some of his property as repayment. Not everyone felt that they could manage such a risk. Yet Lufti and the others all shared one advantage. Casiavera's reclaimers were almost exclusively planting tobacco on plots in the collective land. They understood that one of the ultimate threats of the moneylender, to take possession of the land, was in this case not a possibility. The rights to the collective land were mediated by the collective council, which had the force of the consensus of the eight thousand people who lived in the community behind it when it came to keeping outsiders from controlling the land. These collective claims provided the individual smallholders a force that outside private rights could not counter.

I caught up with the tobacco "boss" when he arrived to take in a harvest with his pickers. Underneath the canvas canopy covering the bed of his work truck, he sat cross-legged, loose limbed, and casual as a constant stream of calls and text messages passed through his smartphone, gold-rimmed reading glasses hanging from his neck. He watched his three workers—all urban landless, one in his late sixties and stiff-backed, and two younger men—cut tobacco, pile it up, and load it in tall stacks sandwiched between layers of burlap in the covered bed of the truck. The laborers wore flip-flops, and even the younger two had the worn, too skinny look of impoverished laborers.

Right away, the boss complained to me about the typical smallholder crop on the volcano. They don't use chemical fertilizers, the boss told me. Only cow manure, and so the tobacco did not have the smell that it should have. He went on to lament that Aren was the last place he could get people to harvest tobacco around there. "No one else will work it." Perhaps the agent's reputation preceded him. He went on at length about his history, how he came to take over his cousin's tobacco business, but he was reticent to talk about the economics of it or his standing in the community. He told me only that he preferred to use his own laborers to cultivate the crop on land he rented from landholders, if he could for three harvests in a row on the same plot (but no more than three, because after that the soil would be so depleted it would not support another crop), paying $70 or $100 for the land used during each harvest. It was only because smallholders were usually more willing to plant the crop themselves with loans of the needed inputs that the boss offered the contract cropping system to his clients.[8]

In a moment of privacy, Casiavera's village head opened up about the problems with the tobacco contracts. Out by his cattle shed, he told me he worried about the people planting the crop. As their elected representative to the local government, he felt their troubles with the debts that tobacco brought. The boss, the village head told me, was an outsider who disturbed people's well-being. According to him, the boss used his capital and the idea of profits to seduce the less educated and less fortunate in his community into indebtedness. "Once you are grabbed by the boss, with the debt, you will not exit." For both Lufti and the village head, the answer was a turn to individual improvement and initiative. Others who applied their agroecological knowledge were not dependent on the exploitative social relations of tobacco. The tobacco planters should follow their lead, the village head told me, if only they were able to.

Tobacco's troubles are demonstrative of the difficult choices that smallholders face. Even with their reclaiming of the land complete, some of Casiavera's reclaimers found themselves in social relations of debt they felt uncomfortable with, subject to exploitation by a monopsonistic tobacco boss. Those with access to a plot on the collective land but without capital were most likely to enter into these exploitative relations, and not just with tobacco. Other forms of contract farming existed on the collective land.

CRISIS AND CROPPING

Social relations of production, like life, are punctuated by moments of crisis and trial. It is at these inflection points that smallholders most often find themselves locked into a predatory capitalist relation. Take Lufti and the emergency his family faced when his wife needed a cesarean while giving birth to her first child. Both mother and child were healthy, but the family needed to pay the hospital bill. After bringing his young family home from the hospital to the tidy two-story, white-washed house Lufti had built next to one of his family's clove orchards, to pay their hospital bill Lufti promptly entered into an odd arrangement with Irfan, one of the wealthiest men on the volcano. Lufti sold the rights to the clove orchard outside his family's home to Irfan for the rest of Irfan's life in exchange for a lump sum. The agreement also stipulated that Lufti would continue to work this orchard for Irfan in return for a 15 percent share of the clove revenues.

The first few years went well. Then Lufti started a new job for Irfan to plant a one-hectare agroforest up on the collective land. Irfan did not mention any wage payment, so Lufti assumed he would be rewarded with a percentage of this forest's profits, just as they had arranged for the clove orchard. But after Lufti has spent a year planting and tending the agroforest,

Irfan never paid him for the work. Then Irfan shocked Lufti by telling him he needed to stop working the clove orchard and move his family off the land. Lufti believed it was because he had asked to be paid for his labor in Irfan's new agroforest that Irfan told him to leave the clove orchard. Three years after the landholder evicted his family, Lufti was still wondering what would happen with the collective land agroforest plot. "Will he share any of the harvest with me once it starts coming in?" If not, Lufti would be in a perilous position, having lost his family's land and clove revenues, as well as four months of waged labor time on the collective land, and now with a second child.

In an opaque, changing, and unequal commodity relationship, tension between patron and client built. Lufti claimed the landowner was "not brave enough to even look at me. He will not even get out of his car when he passes me on the road." But Lufti said that he himself was the "stupid" (*bodoh*) one. Irfan, the landowner, was the smart one, the educated one, the wealthy one, and so, Lufti told me, Irfan violated his responsibility to share both his knowledge and good fortune with Lufti.

Many folks I talked to about Irfan told me he was a "good" person. Many also mentioned he had a "complete" agroforest growing. And many also added, with disdain, "He doesn't do the work, he pays others." "He is rich." "He has lots of capital. He tells us to work more, but he pays people to work. He does not do the work himself!" A third man told me, "He worked for the New Order government, for Suharto, so he is harsh (*keras*). He does well for himself, but for others, I don't know." Lufti claimed that people in Casiavera were turning away from the landowner now; they did not want to work for him. Only an outcast, a man known as an adulterer and domestic abuser, would work for Irfan. No one else would work for Irfan because he tended to construct predatory relations with those less fortunate than himself, this patron of their community who so clearly didn't concern himself with their well-being,

Even as Casiavera's reclaimers avoided the most predatory patrons in their social orbit and aspired to work for themselves, more conventional informal contract farming remained common. Another wealthy man from Casiavera, Zain, who worked for the provincial planning office, owned twenty cows on the collective land. All of them were tended by other people on their respective plots, who earned a 50 percent share in any of the cow's offspring.

For Zain, it was a beneficent livelihood project. "I loan the cows to my people, but not like a bank, I let the people have the profits. This is social development." Zain used words like "my people," or sometimes "the lowest people," to describe the reclaimers on the collective land.[9] Acquiring livestock required a larger outlay of capital than most reclaimers in Casiavera had access to. I spoke with at least thirty families who had entered into contract cropping relations like Zain offered to try to eventually become livestock owners in their own right. Mansur rented out some of his land to a sharecropper as well. He told me he was too busy with his fruit trading business to tend the land. He too talked about it in terms of his own generosity, helping others by allowing them to use his land.

Mansur, Zain, and Irfan were all differentiated relative to their reclaimer clients as being better-off smallholders who accumulated wealth through access to capital and hired labor. Their clients were also smallholders, but these workers were capital poor and relied on selling their labor and land access to reclaimers like Mansur, Zain, and Irfan in informal agricultural relations that ranged from wage taking, to contract farming and sharecropping.

Close attention to landholdings, incomes, and daily work of Casiavera's residents aids understanding of these differences that unfolded among reclaimers. After eight months of ethnographic research in Casiavera, I created a structured interview that included wide-ranging discussion prompts on topics of agriculture, land access, and agrarian politics. I then invited three Indonesian graduate students and one professor of sociology to live with me in Casiavera for a month to carry out two hundred interviews. There, together we revised the interview prompts and spoke with women and men of adult age in Casiavera using a random sampling technique.

After many months in Casiavera and on the collective land, I had come to see differences among Casiavera's residents as key to understanding the ways their reclaiming was able to shape a collectivist smallholder economy. Accordingly, we grouped our interview respondents into three differentiated groups in relation to the collective land: people who lived on the collective land, people who did not live on the collective land but cultivated a plot on it, and people who had no access to the collective land at all.

We found that some twenty families had gained permission from the council to construct what were to be temporary wood homes on the collective land because they had no other land to live on, in return for paying a $2.50 yearly lease fee. Most of these simple homes lacked the televisions,

TABLE 4   COMMUNITY LANDHOLDINGS, INCOME, AND DEBT

Relation to Collective Land	Mean Landholdings	Mean Income (USD)	Mean Debt (USD)
Live on collective land (n = 23)	0.6 ha	$1,909	$354
Cultivate but do not live on collective land (n = 85)	1.2 ha	$3,476	$175
No access to collective land (n = 91)	2.4 ha	$2,827	$172

Source: Author's structured interviews in Casiavera, 2016.

refrigerators, and gas stoves found in the homes of the smallholders who lived below the collective land, on their own property. Missing in front of the homes on the collective land was the usual collection of motorbikes parked in the yards of Casiavera's property owners. Many of the homesteaders living on the collective land did not often come down to the mosque or to attend weddings, hesitant in their status as the poorest of Casiavera. They often wore shorts, shirts, and hats that were handouts from political parties, nothing like the elaborate batiks trimmed with gold thread the wealthier wore. Still, their one-quarter hectare plots provided for basic survival. This was to be the purpose of the collective land, to provide a social safety net to Casiavera's most marginalized.

The families who lived on the collective land earned on average about $2,000 a year (see table 4). This average income was almost three times greater than the incomes of the roughly fifty million Indonesian families who lived on $700 a year or less, the World Bank's defined poverty line for the nation.[10] Even though the reclaimers who lived on the collective land identified as the least fortunate living in Casiavera, their lives were far from the severe poverty that Indonesians who live at or below this poverty line—many of them urbanites—experience. But in their self-portrayals as being among the poorest in Casiavera, these reclaimers were correct.

Those living on the collective land earned a little more than half of the average income of the group of Casiavera's smallholders who had access to a plot in the collective land but lived down below, in the village and not in the erstwhile plantation. That this second group enjoyed a larger average yearly income (~$3,500) is not surprising. These middle-class workers did not live on the collective land because they had access to their own family's land in addition to a plot on the collective land. It is likely that this group, especially, was wealthier and more influential than most in Casiavera already before reclaiming began, because they were the workers who gained access to the collective land even though they already had land elsewhere. Their typical incomes placed households in this second group in Indonesia's middle class; the mean income in Indonesia in 2015 was roughly $4,000.

The third group, those who did not have any access to the collective land at all, had an average income ($2,800) squarely between the average incomes of those who lived on the collective land and those who used the collective land but had land and homes below in the village. Of the three groups, these people were the least likely to be smallholder farmers. Many spent most of their time as construction workers, merchants, traders, teachers, and bureaucrats. As such, there was a large variation in incomes in this group, from landless agricultural workers at the low end to a few construction and food industry small business owners at the upper. The fact that the office workers and entrepreneurs of this group had incomes in roughly the same range as many of the reclaimers engaged in smallholder livelihoods on the collective land supports the widespread belief among reclaimers themselves that a collectivist, smallholder economy was a feasible foundation for a middle-class life.

Reflecting on this differentiated agrarian milieu, what was most surprising to me was that the people who lived on the collective land were able to gain access to the land at all, given their lower wealth. Repeatedly, reclaimers and family-lineage leaders told me that the poorest in the community were given priority over other wealthier Casiavera residents to access the land. The evidence from my structured interviews bore this out.

A similar picture emerged from measures of these three groups' access to land. Those who lived on the collective land reported that they had access to the smallest amount of land (~0.5 hectare); the smallholders who used the collective land but did not live on it reported a larger average value (~1.2 hectares), more than twice that of the people who lived on the collective land. People without access to the collective land at all had twice the average land holdings (~2.4 hectares) as those with access. That this group, richer in both land and income, did not have access to the collective land speaks volumes about Casiavera's residents' commitments to their reclaiming as a movement of agrarian justice.

Taking these three groups as sociological categories does in part obscure important differences within these groups, but it gave me traction in understanding how the collective land was used. It is evident that the land was used by two different kinds of women and men, those who lived on it and were otherwise landless, and those who did not live on it and were part of Casiavera's landholding middle class.

Even while the collective land became a resource of survival for some of Casiavera's most impoverished, it was also a source of accumulation for a few of its most wealthy. At least three families gained control of more than the one-quarter hectare of collective land they were entitled to. At least three families held access to more than one hectare each, acquired through shadow transfers among extended family. While still middle-class landholdings, they were many times the allowed limit. The village head was aware of the land transfers and considered them a problem. "That is not allowed. Later it will become a danger for us!" But he seemed unable to stop them.

The over-the-limit landholders were without exception members of standing in the community. They produced commodity crops, mostly mixed banana and cacao agroforests, and elephant grass to feed their cows, which they housed on the land. Most paid laborers, often within their own families, to tend the plots. All these elites also held land below in the village under various forms of private and family-lineage ownership. The largest landholding family claimed access to nine plots around Casiavera and one plot in the collective land, totaling around six hectares. The family hired labor to work more than four hectares of this land under tobacco. These wealthiest of the community used their plots in the collective land to accumulate still more wealth and move further ahead, while the poorest became trapped in the simple reproduction squeeze, trying to wring ever more yield from their small plots shared by more and more family members.

Focusing in again on these, the poorest of the reclaimers, those that lived on the collective land, brings attention to the fact that after twenty years of reclaimer control, this group was still able to access plots on the collective land. The differentiation that always accompanies commodity cropping did not culminate in the dispossession of the already most marginalized in the community. To avoid their ejection into the ranks of Sumatra's millions of landless poor, the group living on the collective land sought out wage labor more often than Casiavera's other residents (see table 5).

Cash poor and with small farm plots, nearly one-half of the people living on the collective land reported selling their own labor in return for an hourly or daily wage in the month before the structured interview. These reclaimers

TABLE 5   AGRARIAN MOBILIZATION AND RURAL WORK

	Rate of Participation by Labor Activity (%)			
Relation to Collective Land	Union Organizing	Cooperatives	Reciprocal Labor Groups	Wage Labor
Live on collective land (n = 23)	24	55	47	45
Cultivate but do not live on collective land (n = 85)	18	67	29	30
No access to collective land (n = 91)	10	34	26	28

Source: Author's structured interviews in Casiavera, 2016.

also, on average, had debts twice as large as the rest of the village. At the same time, reclaimers living on the collective land participated in forms of collective work at a rate roughly equivalent to their wage laboring. Reciprocal work groups and labor exchanges in their collective land plots were just as common as wage laboring, with many people actively engaged in both activities in a single month. Wage laboring gave them cash in their pockets, for immediate needs. Collective work sustained their efforts to plant new plots, tend their agroforests, and bring in harvests.

Busy as they were laboring for cash while they worked to establish their agroforests, about a quarter of those living on the land also found time to be active members in a smallholder union. This was double the union participation rate of Casiavera's residents who did not live on the collective land but had a plot on it. Casiavera's landless union activism suggests the importance of the reclaiming movement for those who depended upon it the most; those landless who were most exposed to the potentially exploitative aspects of commodity cropping were the most likely to seek the mutual aid and solidarity found in agrarian movement organizing.

Differentiated in their interest in union activism, wealth, and relations to the land, Casiavera's reclaimers replaced the Dona Company's rigid wage laboring (i.e., "coolie") with a variety of more fluid forms of work. Some of these social relations are more favorable to individual reclaimers than others. Casiavera's poorest group, those who were landless laborers at the start of their reclaiming, were the most likely to find themselves involved in predatory relations of commodity cropping.

It makes sense that differentiated relations of production would reemerge as smallholders reclaimed the land. Relations like contract farming were the result of reclaimers having regained access to land but being unable to make it productive themselves. The patron provided capital, the client land. It also signals the monopsonistic control traders had over some of Casiavera's smallholders. Access to many commodity markets was constrained by no more than two or three powerful capitalists. Sharecropping meanwhile signals that even with their reclaiming of the collective land, some residents of Casiavera still did not have access to enough land to cultivate it as autonomous producers.[11] These forces undoubtedly reflect increasing class differentiation in Casiavera, and so, in the long run, they threatened the council's vision of using the collective land to slow processes of dispossession from below.

Exploitation of reclaimers continued in Casiavera. Not everyone was moving forward along trajectories of autonomy and well-being.[12] Even while Casiavera's reclaiming movement enacted new forms of property and territory, those who were most marginalized remained in danger of becoming ever more impoverished. Yet for all the challenges and exploitation, the reclaimers that remained on the land sharpened their commitment to the collective land and a smallholder economy that provided them a chance at well-being. Even as they grew wearier, these reclaimers continued creating ever more economic and movement relations to sustain their individual, familial, and collective benefit.

CHAPTER 8

Reclaiming Solidarities

Even with the hardships and differentiation of the smallholder commodity economy, Casiavera's reclaiming nevertheless held. On their collective land Casiavera's reclaimers found a measure of autonomy that they did not enjoy as agricultural laborers or construction workers. It was not the material properties of their plots or cultivars that made the biggest difference, but instead a situated history and political economy. Reclaiming nurtured solidarities founded on ideas of smallholder autonomy and egalitarianism. Reclaimers acted upon these solidarities with their movement organizing, creation of cooperatives, and trading networks. These social relationships, practices, and movement institutions through which reclaimers expressed their mobilization and livelihoods mattered a great deal.

Late in the 2015 dry season on the Aren volcano, one morning turned from mundane to extraordinary when Henry Saragih, founding member of the Indonesian Peasant Union, the largest smallholders' union in the archipelago, arrived. One of my better friends on Aren volcano, Malin, let me know Saragih was on the way to visit him in Casiavera and invited me to join their discussion. I arrived just as Saragih and two younger men from SPI pulled up along the muddy track in front of Malin's one story, seven-room home. By then, after five months living in Casiavera, I already knew this was not the first time Saragih had come to Casiavera to see how things were going. During my time on the volcano SPI members told me that Saragih's work in the community extended back for more than fifteen years.

Even before SPI's founding in 1998, Saragih had worked with a still-underground local peasant union, the West Sumatra Peasant Union (Serikat Petani Sumatera Barat), to support Casiavera's reclaimers. Along the way, Saragih and a board member of the West Sumatran peasant union worked to get twelve smallholder cooperatives started in planting organic vegetables and tree crops on the reclaimed land.[1] The SPI cooperative members shared techniques to rehabilitate the ruined plantation soils, plant trees, and make organic fertilizers and pesticides, building on revitalized conceptions of organic agriculture that emerged in the region in the 1990s.[2]

While the cooperatives spent long hours cultivating the land, Saragih and other full-time peasant union staff members spent years working across a strengthening network of activists and lawyers in Sumatra and Java to elaborate a legal argument for Casiavera's smallholder land control. Most central of these organizations in the legal and national political aspects of the struggle was the West Sumatran Research and Advocacy Institute (Lembaga Riset dan Advokasi), founded in 1996. A group of former student activists built the organization, at great risk to their own safety, into a group of organizers and lawyers with expertise in democracy and human rights campaigning. Over a number of years of interviews and consultations, the institute came to the view that Casiavera's reclaiming was justified because the original dispossession of the land involved intimidation and violence, and as such the land was stolen from the residents of Casiavera in violation of Minangkabau customary law.[3]

Among the group present that day of Saragih's visit was a construction worker who applied SPI's agroecological techniques in his tiny home garden on the days he could not find work, a government agricultural extensionist, a sugar palm tapper, the manager of a nearby mosque, a fruit and vegetable trader who had recently kicked his meth addiction, a few smallholders who lived up on the collective land, and three movement organization leaders who traveled there to meet Saragih. Most of them also once worked as laborers on the Dona plantation.

Malin, the activist, smallholder, and retired bureaucrat who had first shown me around Casiavera, passed out small glasses of coffee and milk. He told the group that he produced the milk himself, up on Casiavera's collective land, and his friend cultivated the coffee over on the other side of the volcano. Both the coffee and milk, he told us, were organic and a "true peasant product." Once Casiavera's reclamation seemed more or less secure from the return of agribusiness, SPI provided Malin with a bit of capital to get the dairy cooperative going. Saragih had been paying close attention to Malin's project ever since, wondering if it would provide profits enough to justify the never-ending morning work up at the cattle shed required to keep a dairy running. As Malin served us his coffee, the room filled with the retelling of moments the activists had shared during the early years of Sumatra's reclaiming movement. The conversation flowed from reflection to analysis of the current goings on at the Casiavera reclamation, then on to ideas for new potential reclaiming sites nearby.

We chose to gather at Malin's home because no one had been more closely involved in reclaiming the land than he. Born in Casiavera fifty-five years earlier, Malin anchored SPI's efforts to support his village's reclaiming, starting first as a young university graduate under the New Order. Malin came of age as Indonesia's own form of the Green Revolution pushed small-holders into using hybrid rice seeds, chemical fertilizers, and toxic pesti-cides. "We were forced to use chemicals [for agriculture], to use this poison," Malin told me.[4] He joined the land occupation because "people have broken the land, and so now we look for a path out."

For Malin, the path to a smallholder livelihood in Casiavera started with SPI study groups. Malin saw learning and self-improvement as critical to constructing a path free from the plantation:

> With SPI there were changes. We brought the idea of developing ourselves, our human resources. We did study groups, I picked a few people, and we studied together. The start, what brought people to me, was agribusiness. As they learned they could draw comparisons, the people could think for themselves.[5]

These discussions eventually led to meetings with other SPI groups in the province and Malin's network of government agricultural extensionists, which began to develop an expertise in agroecology. After Malin joined others from Casiavera to visit an SPI center for learning and practice focused on agroecology in Java, he helped set one up on Casiavera's col-lective land. The center functioned for the first decade of the reclamation as a place for reclaimers across Sumatra to connect and learn together, or per Indonesia's agrarian movement tradition: "To study together. Ques-tion together. Work together. Everyone is a teacher. The universe is our school."

As in other additional SPI reclaiming movements I have visited, alle-giances were formed primarily among landless laborers.[6] The majority of Sumatra's reclaiming movements include people with a colorful mix of many life histories, brought into relation with one another through the emotional experience of participating in the act of land reclaiming itself. At reclaiming sites, Sumatran-born rural workers join migrant laborers and local small-holders who work as waged laborers as well as sharecroppers and autono-mous producers—all within a lifetime or even simultaneously across diverse and often distant locales. In this way reclaiming movements incorporate both landless people striving to create new connections with the land as well as smallholders who have recently lost their rights to land or are defending them from impending dispossession.

Malin and a handful of his longtime friends made up the core of the peasant union in the early days of reclaiming. They held meetings with their neighbors and interested people from other villages and towns on the vol-cano. The group sought out Casiavera's family-lineage leaders for discussions about taking back the land. Peasant union members were among the first reclaimers to actually go up on the land to occupy it. SPI's numbers grew from those early days to eventually include about one hundred people from Casiavera, mostly married couples and single men.

While SPI members never included even 5 percent of Casiavera's pop-ulation, they made up many of the early reclaimers and so were able to shape the occupation and use of the collective land. As time passed and the tasks before the reclaimers shifted from activism to cultivation, twelve SPI reclaimers' cooperatives drew in more members. Malin and the others at the SPI learning center became the bridge that joined SPI with the Casiavera Council. It was primarily Malin, not Saragih, who made the union mem-bers' messages known to the council.

Lufti was drawn to the SPI center and smallholder collectives. A few years after the occupation began, after spending his early twenties driving a logging truck across the Bukit Barisan for one of the world's largest timber pulp producers, Lufti returned to Casiavera and joined up with Malin's SPI study group. His studies brought him in touch with other reclaimers, farm-ers, politicians, and government agricultural extensionists. Eventually Lufti became one of the most energetic forest farmers on the volcano. Across his mother's and wife's land, he planted hundreds of clove, rubber, cacao, lime, mahogany, papaya, and sugar palm saplings.

Early on in my fieldwork, on his wife's family's land just below the reclaimed land, Lufti told me about the learning and organizing with SPI that allowed Casiavera's residents to constitute an agroecological politics:

> I joined up through those study groups. . . . SPI made things go smoothly in Casiavera, they brought energy. They told us that this was a good place for agriculture, they made us believe we would have this land, we would hold onto this land, so we had enthusiasm. We saw that we had so much support. SPI just made things clear.
>
> We have this land together, we alone could not get it, but they have that group, what is it called? With all the countries, thirty countries . . . [Via Campesina? I interject]. Yes, that is it. They have people from all over the world, telling us we can do this. How can we get evicted (*digusur*) again if we have support from all over the world?[7]

Lufti began that day's comments with the importance of establishing a smallholder economy: an agroecological economy that frees smallholders from debt relations and a reliance on industrial agriculture's debts, inputs, global commodity markets, and ecological paucity. Lufti believed that even with all of SPI's flaws and contradictions, economic change was only possible with the support of SPI organizing, membership numbers, and the fact that it draws power from a coalition that expands across the world.

Here is the linkage of two concepts, smallholder economy and collective social mobilization, that distinguishes reclaiming as what Richard Peet and Michael Watts call a liberation ecology.[8] The integration of collectivist agroecological techniques and social mobilization into a unified field of thought and practice was a pragmatic ideological decision, one that has proven especially politically and ecologically powerful.

SPI members came to understand that their lives required collectivities. Many took to telling those around them that life is "heavy to bear together, light to carry together" (*Berat sama di pikul, ringan sama di jinjing*). They started one of their first cooperatives in 2002, the Wake-Up Smallholders Group (Kelompok Tani Bangun) with ten or fifteen members, after they gained a plot in the collective land from the Community Council.

It was a place where members discussed people's rights, the land, and how they would use and protect it, members Malin, Daud, and Aren all told me. Aren had often visited, as the provincial SPI representative. They focused their early efforts on perfecting an organic fertilizer they could make from plants and dung they found on the collective land, to support the early reclaimers' cultivation. Daud told me it may sound like a small thing, but many reclaimers had the idea that they could only farm with fertilizer, and they were just waiting around for the government to send them some, which was never going to happen during the occupation.[9] Then the cooperative tried cultivating vegetables, because the council was adamant that they should not plant tree crops. They did not have much success in terms of profits, but they learned much about keeping a cooperative going, Daud told me, and realized that this knowledge alone was valuable, so every month they met with other cooperatives to share their experiences.

Susilo was another early member of the cooperative. He took the lead in writing a proposal and filling out the required forms for a government grant to support their experimentation in organic vegetable farming. The group thought government funding would legitimize the occupation. Their proposal was funded with about $1,000. They planted greens, carrots, eggplants, peanuts, and other vegetables. The government gave them a certificate of organic production and a signboard to put up on their plot on the collective land. But the group never did make any money, and Susilo was left with the feeling that they did not get much out of working with the government.

Beyond the money, reclaimers sought state support to learn more about marketing, transport, and markets. But Daud told me, "The government studied from us, we knew more about these things than they did!" This realization led the group to change their focus to become a training center in agroecology. They built a functioning demonstration farmstead and a small dormitory and classroom on their plot. It became a place of learning and support for some fifteen cooperatives that reclaimers would establish on the collective land in the following decade. Building on insights from the Wake-Up Smallholder Group, these cooperatives turned to tree planting, dairy, and livestock efforts on their respective plots on the collective land. Along with working the land, they built connections with trusted movement traders and buyers in the closest urban centers around them.

———————

After the coffee with milk, we walked the collective land, planted in tens of thousands of fruit and timber trees, just below the verdant cloud forests higher up on the volcano. We stopped at a cattle shed, where a cooperative called Sustainable Ambassador was raising cows endemic to Bali.

Saragih mentioned he had just returned from Brazil, where he had visited with the Landless Workers' Movement (Movimento dos Trabalhadores Sem Terra, MST). There, he studied MST strategies of land occupation and agroecology.[10] Saragih told the group that in the arid Cerrado grasslands where most MST settlements are located, smallholders must graze their cows over vast stretches of land. By way of comparison to the MST's grazing, the cooperative members had run through the amount of land their cattle operation needed to grow fodder and their profits over the previous year on the reclaimed land. The cooperative had sixteen cows. Members grew small patches of elephant grass for feed on their individual plots, totaling about four hectares of land. With one-quarter hectare of land, each cooperative member could sell one Balinese cow for $1,000 every three years. Saragih's eyes lit up: "That is more than oil palm." Saragih was always looking for smallholders who found other ways to make equal or greater incomes than oil palm, Sumatra's most controversial agribusiness commodity crop. His enthusiasm was infectious. In the cattle shed there was talk of a smallholder life on the reclaimed land where, with dedication and a bit of luck, reclaimers' children would be able to afford to attend university. They might even go on the hajj in their retirement.

For one of its founders, Luzon, the livestock cooperative was the culmination of a decades-long struggle to get the land back and make good use of it. It wasn't until 2012 that this cooperative came together. Luzon wasn't an SPI member, even though other founders of the cooperative were, because he had long tried to win local political office (without success) and thought the peasant union too radical for his persona. But he was grateful for SPI's role in the occupation and for working on ways to make productive use of the land, Luzon told me. With fifteen members each tending two or three cows, the cooperative is but a tiny operation in the scale of Indonesia's cattle industry, but it serves to provide a place for part-time, flexible livelihood without giving up what Luzon told me was his motivation, to avoid "not being able to work on our own."[11]

One of the younger SPI members, Mansur, had worked with Luzon for a while with an earlier, now disbanded, cooperative that planted mostly bananas. Luzon bought his first laptop to write the proposal and fill out the forms for a grant from the Ministry of Agriculture. Its twenty-five members, mostly men, were given fertilizers, agrichemicals, and a banana cultivar favored by buyers in Medan, Sumatra's largest city. The problem, both

Luzon and Mansur told me, was that they didn't have any knowledge of the land, so they accepted the fertilizers and pesticides the government gave them. But then, when the government stopped giving them these inputs, "We could not work on our own."

A few years later Mansur helped start another cooperative, planting mixed agroforests, mostly other fruit trees, but a few bananas as well. Within a few months a trader had added a weekly stop on their plot to his route, to buy their harvests for sale in Medan. The cooperative grew to nearly thirty mostly younger men and women, so they had two plots in the collective land. "The key was that we overcame the idea that you just send in a proposal to the government to get aid. We don't get aid. We are not even a government cooperative," one of the cooperative members told me.[12]

Reclaimers' desires to disrupt the cycles of smallholder-state dependency that they observed with formalized cooperatives led them to forgo the chance of government funding and inputs, instead creating scores more of these "autonomous" or informal, small-scale cooperatives on the land not accountable to the state. Malin elaborated on the agrarian ideology underpinning the formation of these cooperatives: "We understand a [formalized] cooperative as government-sanctioned organization that is regulated to act strictly as state accomplice. We are not that. We are a smallholder group because the government has become only about money. We are about brotherhood and solidarity. We are humanists."[13]

The integration of ideas and practices related to nature, work, and agrarian capitalism that exist on the collective land can also be found in what are now transnational reclaiming movements across Indonesia and beyond, into other places where rural workers and their landscapes are changing as rural capitalism expands, runs its course, falters, and begins again in new form. Casiavera's workers' reclaiming of the land and forest farms is a component of an egalitarian smallholder society that agrarian justice movements are working to bring to millions of others. SPI's training center in Casiavera, especially, became a center of SPI's West Sumatran provincial organizing, with hundreds of members visiting over the first decade of the reclamation. With work like this, Saragih and the others in SPI sustained steady member movement, from Suharto-era underground safehouses and stints in jail; to meetings in the homes of SPI leaders and their respective communities; to street marches, international climate negotiations, and visits with Evo Morales in Bolivia and Nicholas Maduro in Venezuela.

WORKING FOR AUTONOMY

I first met Sino in 2015, when I was out on a walk across the collective land. Sino was finishing up planting a few clove saplings. He invited me to join him under the shade of his work hut. Sitting with me, Sino pointed out the one hundred or so cloves in front of us that he had planted in the shade of his short banana palms. After the clove trees took hold, Sino told me, he would plant avocado trees.

In his early forties, Sino came to work this land late in life. After ten years working in a paper factory in the nearest big city, Pekanbaru, Sino and his wife Citra moved back to Casiavera, where Citra was born, after they got word that their family and neighbors had gone up on the old plantation and occupied it. Sino and Citra did not participate in the early years of the movement's land occupations, protests, and blockades. Neither did they become SPI members. Yet they were also reclaimers. Because Citra was from there and her community continued its centuries-long practice of upholding matrilineal rights to land, the Casiavera Community Council recognized their need for land and gave them a half-hectare plot in the highest reaches of the old plantation.[14]

As Sino and Citra turned over the earth, collected seeds, planted a diverse collection of plants, and tended them, they were carrying out the work of reclaiming. And with their work they were changing. By working first as landless laborers and then smallholders, Sino and Citra cut against the grain of the most dramatic social change capitalism has wrought across the landscape: the recasting of smallholder farmers as waged laborers in the factories and plantations. When the couple moved their family to Casiavera, they were turning away from their previous agrarian work laboring in an oil palm plantation and logging operation, as well as the time they had spent living in the city earning wages in a sweatshop and paper factory.

Late in the afternoon that day Sino and I sat quietly, watching as a mist flowed out of the forest above and down the flanks of the volcano to envelop us. Our conversation about the potential for cloves and banana palms was interrupted when Sino nodded at a sugar tapper setting out into the forest, carrying a few bamboo containers used to collect sap from the sugar palms that are scattered across the farmland and cloud forests on the volcano: "It makes me happy to see people like Hidayat here. He stands on his own. He is making it himself." Sino, as was the case for many of the reclaimers

I came to know in Casiavera, did not talk about well-being as something defined by survival, or wealth, but instead by autonomy, "making it yourself" or the freedom to work how and when one chooses. Looking out at Sino and Citra's new farm, it was not hard to understand why. Both Sino and Citra had labored for more than a decade in a handful of difficult jobs out in Sumatra's countryside of rapidly expanding logging estates and plantations. Life was indeed precarious; their family lived in what Sino called a "day to day" kind of poverty.

As a young man in the 1990s, Sino worked as an illegal logger for a "boss" in Sumatra's Bukit Barisan mountains. He started as a timber spotter out in the forests, walking in search of the most valuable timber trees. He worked his way up to using the chainsaw, then drove the ten-wheeled logging truck. According to Sino, it was not "official" (resmi), but nor was it completely illegal. The operation exported logs to Shanghai through a busy port on Batam Island. Sino told me, "It was plunder. We took by force (merampas)." This was late in General Suharto's New Order authoritarian regime. For five years he worked the forests. But Sino explained, "I became skinny while the boss became fat." After Suharto fell and Indonesia entered the democratic Reformasi era, the police started to become more of a real threat, arresting entire logging crews. Inspired by the changes sweeping the nation, Sino went in search of some other way of making a living.

Sino floated around for a while before he found work in an oil palm plantation near where he was born, farther south in Sumatra's Bukit Barisan range. His job on the plantation was to cut the fruit bunches at the tops of the palm trees. In a typical day he could harvest one ton of oil palm fruit and load it onto company trucks. The plantation company paid him the equivalent of $10 a day. The weeders, pesticide sprayers, and fertilizer spreaders working alongside him in the plantation would get closer to $6 a day because they were typically women, and their managers, almost exclusively men, said their work was not as physically demanding as cutting the oil palm fruit. It was more dangerous, however, because of their higher exposure to pesticides.

Sino could not bear (tidak tahan) laboring in the plantation for long. "Why did the forest clearing have to be so wide, so expansive?" Sino witnessed industrial agriculture's transformation of the landscape. "The rivers ran dry. The rhino disappeared." The boredom was the worst part, and the persistent surveillance by the overseers (mandur). Sino moved again, this time to Riau in central Sumatra, where industrial forest exploitation provides work. A friend got him hired on at the Indah Kiat timber pulp mill, in the city of Pekanbaru, one of the world's largest producers of cheap cardboard and paper. Sino started as an assistant in the transportation division. After a decade, he worked his way up to driving a forklift, shuttling shipments of chemicals used for pressing the timber pulp into paper fiber.

After having no land at all, the couple often spoke with me about how now they thought their fortunes had improved at Casiavera's reclamation. But when he spoke about his life spent laboring in the plantations and factories, Sino's voice switched registers. "For twelve years, I worked in Indah Kiat. I never got my freedom. It was the same working on the oil palm plantation."

With their plot in Casiavera's reclaimed land came an opportunity beyond what Citra and Sino had found in the palm oil plantations or pulp and paper factory. When I was chatting with Citra, she told me that she was humbled when she, Sino, and their one-year-old son returned from the city broke:

> We are not going back to the city. The factory boss, he controlled us. Sometimes, there was not enough to eat, and we bathed in dirty water. To wash, we walked through our neighborhood to the river full of trash.

Sino too remarked on the improvement and freedom his family had found on the volcano:

> It is calmer (*tenang*) here. We are free (*merdeka*). I work for myself. My work is my own creation. There are not all these rules for everything here. And not so much wealth everywhere.
>
> I get to enjoy life myself (*gemaran sendiri*). My life is refreshing (*sejuk*). Even if I never earn more than ten dollars a day, I get to do it for myself. And I am surrounded by people who may be poor, but who value the people around them . . .

Citra and Sino juxtaposed laboring in factories and plantations with small-holder agriculture as a matter of freedom, aesthetics, and morality. For them, working on the collective land brought the rewards of individual initiative in establishing a livelihood. They enjoyed the friendships and the rural atmosphere. They felt free in a way that they had not before, when working under overseers and bosses.

On one of the many hot, slow days I spent with Susilo as he went about the collective land, he reflected on the landscape around him, telling me things he knew I already knew, but wanting to make sure I would understand their significance:

> With this we live, construct houses, and send our children to university. The day rate of a laborer is not enough to do that. Our lives are hard, and in a way minimal, but we are free! Free to live.[15]

Susilo's thoughts resonated with Sino's, Lufti's, Malin's, Wangga's, Tia's, and so many others' comments. For the first time, as their farms took form, the reclaimers could see their path from rural laborers into a smallholder livelihood in front of them.

Their creation of specific forms of smallholder commodity relations is what gave them their sense of freedom. These were not the forms of subsistence that only occasionally come into relation with circuits of capital.[16] Casiavera's reclaimers were commodity producers. Agricultural production on the collective land was connected to capital circuits, primarily through the sale of fruit, vegetables, cattle, and timber crops to a Sumatran center of industry, Pekanbaru, five hours to the west across the Bukit Barisan. Susilo, Wangga, Lufti, and others created commodity relations as the economic basis of reclaiming. As such, they worked hard to shape and pattern these relations of commodity production and sale into part of their reclaiming movements' working solidarities.

COLLECTIVE WORK ON THE LAND

In 2016 reclaimers worked in fourteen agricultural cooperatives, bringing together hundreds of newfound smallholders to help produce and market milk, livestock, bananas, cloves, cinnamon, avocados, and many other high-value tree crops. Creating collective uses of the land created interwoven social relations, expressed in these cooperatives. They were institutions that emerged from the land and reclaimers' connections to it that attended to planting, cultivating, harvesting, and trading crops in ways that were not atomized like the social relations of production of large-scale corporate agribusiness are. Smallholding, especially the kinds of diversified, agroecological smallholding created in Casiavera, was a socially activating activity, in which cooperatives and casual work exchanges were necessary to sustain ecological assemblages and smallholder economies.

Joining the SPI dairy and two livestock cooperatives were many more cooperatives less affiliated with any single peasant union, but still infused with reclaimers' collectivities. There was the Flamboyant Smallholder Women's Group (Kelompok Tani Wanita Flamboyan), founded in 2002 immediately following the delineation of the plots in the collective land. In 2003, after the cooperative registered with the local Ministry of Agriculture office, the governor of West Sumatra came to visit and celebrate their work, strengthening the legitimacy of the still-young reclamation. They planted a range of vegetables alongside tending fruit trees year-round.

In late 2016 the group was just over twenty women. Members were asked to work one day a month with the cooperative, but their work at the time took on an added urgency because drought threatened their families' rice crops. They expected to have to buy most of their rice for the coming year, but they hoped that their fruit tree and garlic crops would still be productive and provide some much-needed additional cash. Tia, the mother I knew to have been one of the bravest protesters during the land occupation conflicts of the late 1990s, was a member of Flamboyant because "work is hard. We work together so we don't feel as tired. So we don't feel alone."[17]

In addition to working their plot on the collective land, the women rotated working in their members' family-lineage lands in and around Casiavera. Building on the deep historical practice of grounding Minangkabau mother-hood in the land, the groups carried on the matriarchate work of exchanging labor in each other's fields. The drought made another component of their cooperative, a no-interest loan program, all the more important as members became ever more in need of cash during this difficult 2016 dry season. The women met every Friday to collect member deposits and perhaps allocate a loan to a cooperative member in need of backup cash for her family.[18]

The diversity of cooperatives' form and function mirrored the variegated uses of the collective land. Members of the Changing Fate Smallholders Group, a group of mostly older men and women reclaimers, thought of themselves as the vanguard of Casiavera's agroecology. Like Flamboyant, they also chose to register with the government and create a formalized, rotating leadership elected every few years. After experimenting with veg-etable and banana planting, the group decided to specialize in the produc-tion and sale of organic fertilizers and biogas made from locally sourced green wastes and livestock dung. The group proved the territorial utility of working with the government in 2015, when their cooperative successfully petitioned the provincial government for approval and funding for paving an asphalt access road up through the entirety of the collective land. Daud told me that the cooperative members, many longtime SPI members, thought it was time to attempt to use the organization to get a paved road built as part of what he called a "clear, true return of rights to the collective land."[19]

Other work groups were less formal, based on family and friend social relations. These tended to be more focused on supporting reclaimers' livelihoods. Some of these work groups were paid: a smallholder contracted her family, friends, and neighbors for a few days' work to plant a vegetable crop or bring in a fruit harvest. Other groups did not pay wages and instead functioned on the principle of reciprocal labor time investment. Reclaimers were more than twice as likely to participate in these collective work exchanges and agricultural cooperatives than their neighbors in Casiavera who did not have access to a plot on the collective land. None of the work exchanges, even those paid in a daily wage, were equivalent to day laboring for a boss. Reclaimers' work was based on long-term relations among themselves, relations that are not at all like those of a migrant day laborer looking for work out in the fields. For generations the daily patterns of Minangkabau matriarchates have rotated around women's work exchange collectives, wherein long-established expectations of mutual aid and assistance serve to limit the atomization of labor and class stratification.[20]

The collective work on the land unfolded as a repudiation of modern dispossession. Rather than relying on any hegemonic capitalist or statist ideology, reclaimers built on their existing Minangkabau matriarchate of women's land control, agroforestry, and work exchange practices. Bringing in ideas from local peasant movements to the matriarchate, reclaimers contributed to a sense of well-being capable of countering dispossessions' alienation, atomization, and exploitation.

TRADING FOR THE MOVEMENT

Collectivist solidarity extended to how the crops produced on the collective land were traded. One of the most active traders in Casiavera was Mansur. Skinny and always wearing cut-off jean shorts, at thirty-three Mansur lived a life of changing chapters. As a seventeen-year-old migrant in the city, selling clothes at a roadside stall, he had been a "five-footer" (*kaki lima*), Mansur told me, because he had only had his own two feet and the three-foot wood tripod he hung his wares on in the streets to survive. He joined the tens of thousands, maybe hundreds of thousands, of sidewalk vendors in the cities. Mansur told me he traveled as far as Aceh. But after a few months he was "set free" by his cousin, who had been fronting him his clothing stock. Then he got a job at a plywood factory in Medan, making flooring. "The pay did not fit" as a laborer, Mansur said.

That is when he became a methamphetamine addict, nearly losing everything. Life was all "sin" and "no-no's" (*pantangan-pantangan*). Mansur told me, "I had to learn that lesson myself, and struggle to stop using, to survive." Then he found hope as an apprentice barber. His boss helped him get clean, letting him stay at the barbershop and not letting him leave in the early days, when he was racked by withdrawal and wanted nothing more than to make his way to the market, where the users hung out and he could find some crystal meth.

Mansur found freedom first not on the land, but as a barber. This work was also how he first learned about the peasant union. After long talks in the barber shop with movement members, he joined up and was quick to hit the streets to protest in the city. When it came time to return home to Casiavera, all the plots in the collective land were already claimed, so Mansur traded fruit grown on the land instead of growing it himself. In the rainy season, February, it was mangosteens, the small purplish-brown fruit with sweet white flesh, that were in season. Mansur sold the fruit to an exporter in Pekanbaru, who mostly sent the mangosteens to Taiwan, where they are popular as both a snack and part of the daily Buddhist offering. A few months later, in May, when I visited Mansur's storage shed built alongside his home, it was avocados that came in as a bumper crop. Mansur told me that for a few weeks twenty women and men had come by every day with avocados to sell.

Another afternoon I watched as six or seven groups of teenage boys brought bags of avocados to his porch over just two hours. That day Mansur was wearing a shirt with the English words "Diversity, Debate and Art" on it. During my visit he ducked through his back door a few times to check in on his pregnant wife. One of the fruit collectors told me this was young man's work, because it is difficult to climb up the trees to harvest the round fruits. In the height of the season, they earned $10 a day. They mostly collected the fruits from the trees scattered around their own family's properties. Mansur weighed the fruit with a dinged-up plastic scale and calculated payments to the young men using a smartphone. The going rate changed

FIGURE 18. Mansur the movement trader sorting avocados at his home.

every day, based on what he was paid the previous time the exporter had arrived at his home to buy the crop. Although Mansur graded the fruit into levels of quality and was paid differently for these, Mansur told me that he would not allow the pickers to grade their own to receive differentiated prices: "It leads to too much arguing." Mansur was paying around fifty cents a kilogram at the time. There were no negotiations.

The way the pickers dropped off their crop and received their payment without any negotiation was only possible because they were closely tied together in a relationship of reciprocal obligations. Both parties lived in the community; many were family. They were embedded within a durable web of relations. The pickers trusted Mansur to give them a fair price, even if they did not know what the going market price for mangosteens or avocados was or how much Mansur would be paid for them. The transactions seemed very different than what a micro-economist would consider rational economic decision-making. It was instead a transaction based on trust and a sense of common benefit.

When talking about his business, Mansur was careful to encourage these perceptions. Mansur used words like "we," "help," and "work together" to define his position as a trader. He took pains to tell me it was "not just for money, not just for my profit." Still, he and his suppliers were not equal. It was clear from his home that over the years expanded into a compound, his recently opened general store, and his ability to pay the pickers before he was paid for the crop that Mansur made more profits than the typical picker.[21]

Mansur's role as a movement member, cooperative founder, and trader came from his ability to journey into the unknown as a kind of modern urban frontiersman, into a site of danger and adventure—the city—to survive its worst and return, connected with merchants and exporters there. It also stemmed from the thoughtful ways he went about his work as a trader, which increased solidarity and trust among the reclaimers. Mansur's agro-ecological ethic emerged in his concern for others, beyond his own profits. His efforts to support those younger than he illustrated the importance of the concept of autonomy to Mansur and others in the peasant union. "I get the young people learning about trading. I do not keep them as my employees, no! I do not pay them a salary. . . . After they hang around a while and see what I am up to, I will loan them fifty dollars and send them out to see what they can make trading themselves."

Mansur's vision of a thriving economy included a foundation of many independent, small-scale producers and traders. He wanted to go beyond reliance on capitalist bosses to create forms of smallholder economy that offered freedom from laboring for a wage. Even as he wanted to work with the young men around him to gather fruit, he would never consider himself their employer.

Not all of Casiavera's aspiring traders were as adept at this kind of work. Waranto, Mansur's neighbor, spent much of his thirties trying to establish a role as trader of reclaimers' crops. Like Mansur, Waranto left Casiavera as a young man in search of an urban life. His first job was taking tickets on an interstate bus. A year into that he signed up with a government work program to spend three years as a line cook in South Korea and then spent three more working in an electronics factory outside Soul, before returning to Casiavera to start a family with his high school sweetheart. When he returned to Casiavera, Waranto found it hard to work with the reclaimers and was a bit uncomfortable with the ways the peasant union members, especially, went about their work.

Waranto connected with a few of the largest vegetable, fruit, and palm sugar buyers in the closest cities, but reclaimers eventually mostly stopped coming to him to sell their crops. Waranto thought he could not get established as a trader because he spent more of his time getting a cell phone kiosk and corner store going at his house, instead of up on the land and at the peasant union meetings. Now his other small businesses sustained his family, especially the corner store, where he allowed more than fifty reclaimers to shop on credit or trade crops like fruit and palm sugar for their wares. In the few days I spent around Waranto's home and market, these transactions were curt and brief. Unlike at Mansur's place, there was much negotiating over these trades, with both Waranto and smallholders often declaring the "price doesn't fit" and not making a deal for things like plastic boots, fluorescent light bulbs, mangosteens, or palm sugar.

Both Waranto and Mansur talked about their businesses in much the same way. Waranto told me, "I do this for all of us, we all have profits, we share what we know and have learned together until late into the night!" Yet Waranto had not been able to play a role of solidarity and collectivity in the same way as Mansur, and it was easy to feel Waranto's frustration. Speaking about his business, Waranto pointed out that he had "come up against individualism. It is the one with capital that makes profit."

When I asked him about how cooperatives could fit into this kind of capitalism, Waranto made clear that his experiences had left him disillusioned with the organization of the collective land: "Cooperatives? The theory is fine, but really it brings all of us down together. It all looks good, but the peasant unions should ask themselves, why are we still poor if they are supporting us with their cooperatives?" Waranto's life of laboring and now as a small businessperson gave him a vision of the economy as a realm of individualism that didn't include collective work. "What is the solution then? What is the road forward? It is not cooperatives, it is not companies, it is not projects or programs," Waranto told me.[22]

In Casiavera, the communal, cooperative, and reciprocal aspects of life are constantly running up against the capitalist: contract farming and debt. Casiavera's reclaimers may have succeeded in creating a post-plantation place, but it is one in continual remaking. Struggle, negotiation, exploitation, frustration: all these processes are present and in non-stop action.

Patterns of collective work in Casiavera reveal that their reclaiming economy was not a repudiation of capitalism per se. Reclaimers' relations of reciprocity and mutual obligation did serve reclaimers in their desire to distance themselves from certain aspects of capitalist markets, especially any complete reliance on a single agricultural crop or waged laboring. Reclaimers overcame their lack of capital by grabbing back the means of production, the land, and commodifying it with a whole host of labor relations, from wages to agroecological cooperatives. By seeking to avoid debts and agribusiness and maintaining control of the means of production, agroecology acquires its political dimension.

In these efforts of smallholder survival, autonomy is the crux. Motivating reclaiming and collective work are ideas of liberation from exploitations of all kinds, from the plantation system as well as the wealthy neighborhood landlord. By necessity the economic autonomy afforded by a diversified, flexible, collective, and not fully commoditized livelihood takes the center position. This smallholder livelihood is mostly an individual and family effort, to produce primarily commodities, unfolding as the collective ownership of the land to restrict the potential for dispossession to the greatest extent possible. The Landless Workers' Movement calls the kinds of people who carry out this kind of work "free workers" because through land occupation and agroecology they have recovered their ability to create for themselves, instead of living as landless workers in service of capital.[23]

In Casiavera, collectivities of ownership and work gave the most fortunate reclaimers the economic power to produce a mix of commodity crops at their own will, at their own pace, molded in their own vision. An agroecological ethic took form, encouraging complex social ties among reclaimers in which financial transfers were more horizontal and less vertical.

EGALITARIAN TERROIR

Casiavera's residents often spoke about how their smallholder relations were especially "dense" compared to other places. In people's use of the term *dense* they were referencing the strength of their social connections, of how they were not a society fractured along the few vertical lines of industrial capitalism but instead connected through many, more egalitarian, ties. It was not necessarily that Casiavera looked much different than a typical upland Sumatran village in the countryside. It was that many people living there talked about and conceived of their position and purpose within the political economy in a way unusual within Sumatra's dominant trajectories of dispossession and industrial agriculture. As Casiavera's forests—full of a changing collection of species—consolidated, reclaimers became ever more joined through ideology and economy. They did not have the hierarchies and bureaucracies of the state or corporations. The unique set of socio-ecological forces that came to bear on Aren—its terroir, to use the agricultural concept—resulted in an egalitarian slant to society.

Central to this terroir is the reclaimers' existence as a very self-aware counterpoint to the rural workers who find themselves earning wages from international agribusiness, where classist, racialized, and gendered forms of marginalization and abuse have existed since the founding of these companies' plantations. There was a palpable feeling among reclaimers on the volcano that these social ills were not as severe there as they were at the plantations. The feeling that reclaimers' work did not include the extreme marginalization present in Sumatra's plantation lands was strongest among reclaimers who had spent time laboring in industrialized zones, not least those who had labored on the plantation above their homes in Casiavera before the reclaiming movement. Nearly all the smallholders I spoke with objected to the social relations of production through which crops like rubber and oil palm are cultivated on the plantations, specifically the toll on laborers' human bodies and well-being.

When I asked these reclaimers about why they chose a smallholder life instead of other lifeways, many vividly explained their efforts in terms of their desire for a more autonomous day-to-day existence. I quote eight of these peoples' voices, as I and three Indonesian sociologists recorded them during structured interviews:

> "Agribusiness diminishes our power." // "With companies I don't experience any development." // "To not be forced, so that my life is more relaxing." // "So that a company does not dominate our lives." // "So that others can't control me." // "To not be bound by any of their [company] rules." // "So that I am not bound to my work." // "With company work I can't do more than one job at a time." // "Better to cooperate with others rather than work for the company that only thinks of itself." // "It's not enjoyable to be a laborer nor a boss, so why work for a company?"

Many other reclaimers chose to speak about their reasons for joining the reclamation in the affirmative. I quote twenty of these affirmations of the smallholder economy to give due attention to the interconnected issues that make up the concept of reclaiming in the Sumatran countryside. First I quote three voices about the value of the homestead in motivating their desire to be a smallholder:

"To enjoy my own place that I live in, with peace." // "To be closer to my children and husband."// "To leave something, an inheritance to my children."

A few other reclaimers reflected on the flexibility and entrepreneurship of the smallholder livelihood:

"To be free." // "So that I am free to begin and finish my work, I can go home whenever I would like." // "So that I can work anytime." // "To be my own manager." // "My work is not limited; I can earn more money if I work harder." // "To develop with our own hard work." // "If I own land, I can work it—or not—myself." // "Even if our harvests are small, we have freedom and peace." // "So that I do not need to buy anything." // "It is more creative work." // "For my family economy." // "It is more enjoyable." // "To have my full rights." // "To have independence."

Reclaiming politics is not simply an oppositional position. It is a way of thinking about and enacting alternative forms of production—reclaimers' landscapes—that can sustain smallholder economies. This is geopolitical knowledge that is not limited to micro-scale plant interactions.

No one on the volcano could elaborate this reclaiming ecology better than powerfully built and intense Aren, whose parents' decision to name him after the volcano he was born on seemed prescient. Trained as a law-yer, he lived surrounded by rice terraces and sugar palm agroforests and frequently joined Saragih on international delegations as a leader of the province's Indonesian Peasant Union chapter. I first met Aren when Daud, who managed Casiavera's mosque and was one of the first in the village to sign up with the union, took me to meet him. Daud told me I should meet Aren because I was now living in "his area." With a title of indigenous prestige, Datuk Maju Indo, Aren brought influence to the union's grassroots. He ran for provincial parliament in 2014 with a newly formed political party, Nas-Dem. Aren saw an opportunity for reform with the new party and wanted to increase his and the union's influence by entering government. About Aren's campaign, Daud told me, "We put him forward, the people did. So, of course, he did not win."

A winding dirt road brought us around the foot of the volcano and back up a steep river valley to Aren's home. After a curt greeting, Aren took Daud and me back behind his home to a hollow surrounded by rice terraces and groves of coconut and cinnamon trees. A spring bubbled up at its center. We removed our shoes to sit on the cement floor of Aren's prayer hut, which doubled as an open-air movement meeting place. We could talk freely there, in the calm, knowing we were alone. Over the months I would experience Aren's easy humor. But at this first meeting he was much more serious: '

> We gave ourselves power, we entered the borders of their HGU [plantation boundaries]. We took it ourselves. We brought a battle to them. They [the agribusiness] thought we would just be their coolies! They underestimated us. We didn't agree with the idea of a "Land Project" and land being a commodity. We are not here to be a "project" for the world.[24]

As Aren narrated his own form of agroecological politics, freedom is to be found in becoming strong enough to challenge agribusiness's interventionist behavior in Casiavera. For Aren, the state needs only to sympathize with the reclaimers' cause; it need not be replaced. Instead, reclaiming is the end in itself. Land control, not control of the state, is to be the basis of autonomy, much like the way Zapatistas define their movement as one that doesn't seek to capture the state outright, but instead seeks to take autonomy back from it. Casiavera reclaimers join the Zapatistas to create new worlds within their existing societies, "from the bottom and to the left" as the Zapatista motto goes, working toward a pluriverse of worlds autonomous in their ability to defend old practices, transform others, and invent new ones to counter dispossession and capitalist domination.[25]

Aren made clear that he saw capitalist agriculture as an external force, created by corporations with the power to determine his life. In Aren's formulation, the problem started with colonial capitalism: "Capitalism entered the collective land. Not just there [at Casiavera], but all over the region, for plantations, cattle, tea, quinine, rubber, and all those things." Aren thought about capitalism as a collection of external, foreign actors. Smallholder economies were the counter expression, one that for Aren was moral in a way that industrial agriculture, "land projects," and "development" were not.

Out in the prayer hut, Aren drew me a diagram of the power structures that were altering his world. At the bottom were the workers and local government, above them the district head and the governor, then the president and Congress. Above those were the familiar targets of the anti-globalization movement: the World Trade Organization, World Bank, and International Monetary Fund. Finally, at the top of the diagram, were the G8 and the transnational agribusinesses Cargill, Wilmar, and Monsanto. Aren drew a straight line, with each set of actors given an arrow to represent their influence on the ones below. He did not see a system of rapid changes, innovations, and disruptions. He saw a long continuity of capitalist agro-industry, the systematic engineering of the economy by those he believed he would never meet, for their profit alone.

Aren drew me another diagram, this time of five circles, each a bit larger than the previous one. These concentric circles were the land control required to repel the external force of agribusiness, the actor that Aren perceived as bringing capitalism into society. At the center of this second picture was a small circle representing the land a family builds their home on (*tanah pemukiman*). Then came rice and agroforest lands (*tanah sawa dan parak*). Outside of this circle was family-lineage land for agriculture and other uses (*tanah kaum*). Finally, the largest circle was community collective land (*tanah ulayat nagari*). Aren located Casiavera's reclaimed land in the outermost circle as a cultural and economic resource capable of pushing back against dispossession. Although these lands were scattered across the landscape, not actually laid out in concentric circles, Aren's diagram illustrated the cosmological role that land played for him. It is a series of outward-facing moats. They exist to absorb the parries and accumulations of capitalism. If the core is lost, so are the people.

Very old Indigenous Minangkabau conceptualizations of land control underpin Aren's vision of reclaimers' organization of the landscape. Connecting the community, family, and individual is not only collective. It is collective, familial, and individual. The overlapping scales and transmutable social networks of this concept of land control make it difficult to enter from outside of it using capitalist buying and selling. Here it is not capital or capitalist power that makes land and resources available. Instead, it is a cosmopolitan and dynamic kinship created through marriage, community membership ceremonies, and family council consensus that serves to integrate locals and migrants into the long-established Minangkabau family lineages and their forms of customary authority.

Aren's anti-capitalist position reflects that of many modern peasant, landless, and indigenous movements, those workers against global trade, agribusiness, and the finance industry. And like many of the smallholder activists that make up these networks, Aren's agroecological politics goes beyond the familiar litany of "anti-" protest movements to include a way of living and working the land. It was this integration of politics and livelihood that gave Aren a fully formed movement perspective, one that he was able to bring to bear on the difficult situation in Casiavera with considerable success.

Inherent within Aren and other reclaimers' mobilizations is the idea that after they secure the right to the land, agro-industry is to be replaced with landscapes of smallholders, free to produce in the mobile, flexible, and eco-logically attuned ways their world demands. These movements for autonomy are not seeking isolation or autarkic self-sufficiency. Social organization is to facilitate interchange and action between people in other places that have already been ruined or contaminated by industrial extraction. Alliances, dialogue, and connections with other movements are paramount, allowing them to join up to create novel "ways out," as Sherry Ortner has put it.[26]

Aren's politics of a way out from dispossession draws on the rich Suma-tran political philosophy of the Cold War era. Across the volcano small-holders had an interest in the analysis of Karl Marx, Vladmir Lenin, and Mao Zedong. The Sumatran anti-colonialist and communist Tan Malaka carried weight in discussions for the way the political prisoner and scholar integrated critiques of authoritarian capitalism with remembrances of his ancestors' more egalitarian, pre-colonial upland Sumatran matrilineal society. Now, nearly a century on from his theorization of the peasantry in anti-colonial revolution, the needs of smallholders and the landless who desire land are an even more awkward fit with the well-worn Marxist class politics than they were during Malaka's life.

Indeed, Casiavera's reclaimers were not concerned with a consolidated land-owning class; they instead took on the challenge of replacing state-endorsed agribusiness land contracts, licensing agreements, and private security forces. These were structures of power in which class politics had less traction. Neither Lenin nor Karl Kautsky theorized a path for laborers to become landed smallholders.[27] While Marx himself eventually came to believe that collectivist smallholder communities (i.e., "peasant communes") held an element of superiority over capitalism and could provide the "starting point for [agrarian] communist development," the bulk of the Marxist move-ments of the nineteenth and twentieth centuries joined Lenin and Kautsky in denying smallholders and back-to-the-landers emancipatory potential.[28]

Critically, agroecological politics doesn't seek to take collective control of the plantation in the way an old-school Marxist politics would. SPI mem-bers often told me they joined to gain agricultural land of their own so that they could leave the plantation behind or dismantle it outright. Nearly all movement participants had personally learned about the exhaustion, drudgery, and poverty of laboring for agribusiness or on the manufacturers' factory floors. For them, the plantation and the factory were not locales of rewarding and stable livelihood, because their membership in this laborer class did not often provide them a path to well-being.

In many ways, reclaimers' experimentations in egalitarian community making carried through more on communalist-anarchist ideas than Marxist ones. This is a stream of emancipatory workers' politics that harkens back to Peter Kropotkin's 1902 communitarianism, founded on the idea that all humans have a proclivity for cooperation, not conflict. Underlying all anarchist action is the belief that cooperative impulses can sustain self-governing communities where hierarchies and bureaucracies are limited and economies non-capitalist, even as large-scale feudal, state, and corporatist fields of power determine life. A long lineage of theorists of anarchism and communalism has described diverse social forms of this egalitarian, cooperative impulse that exist within the capitalist age.[29]

Certainly the form of ecosocial self-governance that unfolded in Casiavera showed flashes of anarchist social organization. Within the movement were collectivist traditions and a rejection of extreme forms of authority and its economic extension, exploitation: the most basic of all anarchist principles.[30] There were parallels with India's Gandhian anarchism, a movement that eventually threw off its tolerance of the colonial Indian state to strive for a sweeping reconstitution of society with direct action and experiments in repeasantization and deindustrialization. Even so, Casiavera's reclaiming movement looked to other forms of thought outside of Marxist and anarchist thinkers to inform their activist engagements. Class politics gave way to more humanistic and environmentally informed critiques.

In the political conversations I joined in the Bukit Barisan, alongside Indigenous Minangkabau cosmologies, reclaimers often mentioned Jurgen Habermas for the way the philosopher laid out a path toward the humanization of nature, instead of its total subjugation and annihilation. This more benevolent human structuring of the world focused reclaimers' concern with ecology and nature. Consider the utility of Jurgen Habermas's idea that the natural lifeworld is coextensive with society, where one's labor mediates the forms of politics and culture that define the interaction between human and nature.[31] For mobilizations against the plantations and their financiers, ideas about restructuring local lifeworlds as smallholder landscapes guided action toward agroecology and protest in a way that class and labor politics did not.

Reclaiming is an act of recovery, the undoing of dispossession and a reconstitution of something that was lost, but nothing in the past can be completely undone. Reclaiming the land necessarily requires the establishment of a new cultural-political place where original forms of social and ecological production can unfold. Reclaiming is justified by a vision, not of returning to the past, but of moving toward a different future.

On the volcano, reclaimers' emotional register became one of optimistic collaboration to throw off an ecologically disastrous and exploitative form of development. There was a measured excitement in their dairy cooperatives and the upland agroforests taking root where corporate rapaciousness had denied their claims and destroyed the land. These were changes of a more prosperous smallholder economy. Enmeshed in nature, on a spectacularly forested slope of a volcano in the Bukit Barisan, the politics of defining a smallholder territory played out in such a way as to illustrate a path forward, away from dispossession and its polarizing forms of distribution of wealth and ecological decline.

Going Beyond

Can people make something new out of places of dispossession? Is there any potential for reclaiming sites of colonial and capitalist exploitation and ruin? With nearly every mountain, river, and forest in Sumatra's Bukit Barisan mountains now transformed by logging operations, mines, dams, and plantations, the answers to these questions are difficult to see. Most evident are (neo)colonial dispossessions' remakings of landscapes on the largest of planetary scales. At times, claims about the hegemony of capitalism across Indonesia do not seem too unrealistic. Thousands of sites of land dispossession exist. Few of these sites have seen reclaiming movements.

The Casiavera reclaiming movement makes evident, however, that the answer to these questions is unequivocally yes: agrarian peoples can construct ways of living that counter dispossession. Sites of enclosure can be reclaimed. In a place where deeply embedded structural inequalities in land access, wealth, and power existed, reclaimers found a potent movement power in their women-centered, collectivist, agroecological economy. Linking up with a transnational smallholders' movement, reclaimers mobilized to end corporate control of the land and allow individual smallholders to use it for cultivating commodity crops for their own benefit. A vibrant, diverse smallholder ecology transformed the landscape.

The reclaimers' project was oppositional to agribusiness and by extension opposed to international finance capitalism as well. Even so, agricultural livelihoods on the reclaimed land were mostly structured around cultivating commodity crops for sale into domestic food markets. In this way, reclaiming worked to change capitalist agriculture from within it. Enmeshed in a web of capitalist relations, reclaimers worked to alter them in ways that made them more humane and ecologically attuned. Reclaimers' turn to collective land control and collectivist relations of production expands the economic imagination to include ecosocial ways of producing resources that, at their point of production, are not typical capitalist commodities and so are not commensurate with alienated capital or labor.[1] On Casiavera, capitalist relations were shown to be but a limited social force that was only one component of the interactions of "people, labor, sentiments, plants, animals, and life-ways" that bring into being heterogeneous, fragile, and contingent ecologies of both exploitation and emancipation.[2]

The injustices of the New Order sparked the broader Indonesian reclaiming movement, but it was an agroecological politics and the smallholder landscape it engendered that sustained the movement in Casiavera. Most visible across the land are the agroforests that reclaimers cultivated as a way out of the life the Dona Company offered them. Growing underneath Aren's ancient caldera, these valuable agricultural forests are material expressions of politics and ecology. They are a form relating to the land that allows people to live in more mutualistic interaction with many forms of life. Indeed, agroecology and the human-plant relations it depends upon can generate a very authentic affect and attachment.

For hundreds of millions of workers around the world, reclaiming places of dispossession and industrial ruin has become a defining issue, now more than 150 years after the first of the Earth's many industrial revolutions. At the bulk of sites where capitalism has remade the environment—in the mines, logging tracts, factories, and plantations—compounding crises of poverty, ecological destruction, and toxicity make the ecological and social limits of dispossession known and evident. The challenge confronting agrarian peoples in ruined and depleted places is how to conceptualize and nurture ways of relating to the land that state and corporate dispossessions have long repressed. In these places, reconstituting fundamentally non-industrial livelihoods is vital. Casiavera's reclaiming movement brought together notions of agroecology and social mobilization to find a measure of well-being. This well-being, this way of living that made their landscape a more emancipatory one, provides insight and inspiration to other workers who share a vision of becoming autonomous smallholders, but who remain enmeshed in capitalist relations that they consider unjust.

Furthering struggles to counter dominant trajectories of dispossession requires a clear vision of reclaiming movements' current limits. Indonesia gives us the answer to the question of what happens when an authoritarian state hands over the land, nearly wholesale, to corporate exploiters of the land. The answers to this question often involve acquiescence. Facing violence and great uncertainty, many people initially intent on resisting dispossession have found themselves in modes of coexistence with companies. These dispossessed agrarian peoples are the "conscripts of modernity," as Tania Li and Pujo Semedi call the smallholders who, dispossessed by expanding oil palm plantations in Indonesian Borneo, had their old forms of life dismantled. The plantation company obliged these workers to live with the company under exploitative conditions not of their choosing. Their situation became, on the whole, dire. More than a bit of fatalism circulated. Looking out at Indonesia from these landscapes of dispossession, for Li and Semedi the dominance of Indonesia's state-corporate regime is complete; the political conditions for a reclaiming movement capable of systemic change do not exist.[3]

It is true that reclaiming movements are small in scale, and the state continues to take land and hand it over to corporations. Given the vast inequalities in power between agrarian peoples and corporations, this limited influence of reclaiming movements is to be expected. Casiavera's reclaiming is worth celebrating in part precisely because it is something quite difficult to achieve. Even if reclaiming movements are not able to stop dominant global trajectories of dispossession, they work to contain and alter existing fields of domination.

Casiavera's reclaimers, with their sustained involvement in a larger network of reclaiming movements, point to an important reality where resistance and affirmative creation of new livelihoods and landscapes can counter dispossession. These ways of living are already part of the present.[4] Sometimes, alternative forms of socio-ecological organization come into direct confrontation with projects of industrial production. At other times, agroecological life emerges only after the plantation fails, the timber company exhausts its forests, the mine is depleted, or the landscape is otherwise ruined.[5]

Nearly without exception, emerging out of these places is a propitious view of the ability of existing social structures to support diverse agroecological assemblages of life. With original ways of living, using, and relating to the land, reclaiming movements are reconfiguring society. Their ongoing struggles are rearranging plot by plot the relations of agrarian peoples in the economy, across the landscape, and within the state.

RECLAIMING STORIES

Reclaiming dispossessed land presents both a challenge and an opportunity for rural workers. Before illustrating the character and scope of reclaiming as a global phenomenon with a few stories of countering dispossession, it is first necessary to acknowledge that sometimes rural workers must seek liberation in other forms, because in certain colonized places the bar for reclaiming is too high, usually because of its predatory political economy. Take, for instance, California's Central Valley, where recent reclaiming movements have made but the smallest inroads into the region's agribusiness landscapes. The exploitation of the valley's migrant agricultural workers in the corporate-owned plantations is very nearly as complete as one can imagine, criminalized as they are by immigration law, pursued by armed agents of the state, traded by murderous *coyotes*, non-unionized, underemployed, denied affordable health care, often homeless, and continuously exposed to chemical toxins.[6]

In other places reclaiming is feasible, not necessarily because of social movement power but because corporate industry itself abandoned the land. Disconnection from the land does not relieve rural workers from long and contentious reclaiming mobilizations, however. In Indonesia, these contentions are most pressing across the millions of hectares of logged-out, bankrupt industrial logging concessions across Sumatra, Kalimantan, and West Papua. These lands remain a focus of state and agribusiness claims, even after the logging companies abandon them and rural workers construct homesteads within them.

At what reclaiming movement members call the "Hope Forest" in the central Bukit Barisan, for example, I visited a rural workers' occupation of a logging concession the Asialog company had abandoned in 2000. By the time of my visit in 2012, dispossession had taken a new form on the land, with agribusiness's procurement of a state conversion of its industrial logging land concession permit into oil palm plantation and forest rehabilitation business licenses. Returning with the full backing of the state, armed forest police, called SPORC, joined with several other state agencies and a mob of criminal toughs to burn down the workers' Hope Forest homes, bringing reclaiming of the land to a dramatic end.[7]

At other places industrial agriculture's abandonment is more complete, and reclaiming is often more successful. At a bankrupt palm oil plantation in Kalimantan that I visited in 2011, millions of dollars of corporate investment, an oil palm mill, and palm oil tanker trucks had all disappeared. More accurately, they had gone up in smoke after plantation laborers who lived on the company's estates burned down most of the company's buildings and infrastructure in multiple fits of rage. After years of believing the company would return to restart the palm oil mill and begin paying them to work the plantation once again, facing total abandonment, poverty, hopelessness, and many more complicated emotions, the workers took up fire as a tool of subversion and rebirth.

Even while fire was weaponized, it was also regenerative. Workers burned patches of the overgrown oil palms to make space for new plantings of rice. They did so because they could not eat oil palm, nor could they sell it to anyone, because the company had shut its mill years earlier. Amid the fires and devastation, a few Malay families began scavenging what oil palm was still productive. They connected with a palm oil bootlegger, who could sell the oil palm fresh fruit bunches to another mill a few hours away under the table, without a contract. A few more families with a long history of living in the region returned, more emboldened and organized than when they had been evicted from the plantation two generations earlier. They scavenged oil palm too and planted their own forests crops. The government ignored them, and the company avoided them, if it still existed at all; no one was sure. How reclaiming moved forward was largely up to them.[8]

It is not only in Indonesia or on oil palm plantations that the opening and closing of individual zones of dispossession leads to contemporaneous processes of proletarianization and reclaiming within single regions. Take, for example, Central America, where even as the region remains a center of banana plantations, the opening and closing of individual estates includes processes of reclaiming. In United Fruit's shuttered banana plantation properties in Puntarenas, Costa Rica, thousands of landless peoples moved in to carve out smallholder farms as localized counter-trajectories of development. This was after United Fruit had proletarianized many of their ancestors in the nineteenth and twentieth centuries. Where once bananas were produced with tractor, truck, and packing assembly line, the wooden plow and mixed plantings of food and commodity crops dominated again. Proletarians became campesinos. Autonomy begot agricultural production. Diversified agroecological production established a foundation for food sovereignty.[9]

Another site of plantation abandonment, Cuba, emerged as a center of reclaiming. After USSR subsidies for the island's industrial sugar plantations ended, the plantations were largely shuttered. With the kind of state support few reclaiming movements have enjoyed, reclaimers took control of plantations and derelict urban properties to create smallholder farms in the countryside as well as the cities. The new, small-scale farms provided the foundations of an agroecological economy that produced much of the island's food.[10] A state-sponsored union, National Association of Smallholders (La Asociación Nacional de Agricultores Pequeños), aided in organizing and capacity building among a dense network of smallholder cooperatives.

A more revolutionary, national-scale reclaiming took place over the generations following the Mexican Revolution and its dismantling of more than 10,000 latifundia and haciendas. Before the revolution, only 2 percent of the population owned 65 percent of Mexico's land. By 2010 the nation had the largest land area under customary and community management in the Americas. Reclaiming the large holdings of the old regime involved rehabilitating degraded lands, creating new land, forest and watershed organizations, and much smallholder work to construct agroecological alternatives. A regional repeasantization unfolded, even as some smallholders already in place on the land gave up their hoes and sought wage work in the maquiladoras, streets, and across the border. Related repeasantization and smallholder tree crop intensification has unfolded across Latin America, where "insurgent and economically innovative" agroforest farmers increasingly characterize smallholders.[11]

Different kinds of reclaiming are unfolding in the Global North. Detroit is the site of a coalition of city dwellers that reclaimed foreclosed-upon land parcels to create agricultural commons. An elder who began farming first as a member of the Black Panther Party started one of these Detroit reclaiming movement organizations, called Feed'om Freedom. The group grew to include hundreds of active guerrilla gardens across the city.[12] Urban smallholders can be reclaimers too, when they join ideas of social mobilization, agriculture, and control of land to sustain their political aspirations and livelihoods.

Two thousand miles to the west, Sogorea Te', a Chochenyo and Karkin Ohlone women-led reclaiming organization, built a vision of Indigenous sovereignty across the lands of the California Bay Area. Following nearly three centuries of dispossession of the Ohlone peoples from the bay's coastline, Sogorea Te' reclaimed multiple plots of urban land as community farms and centers of learning and organizing. In seeking to restore their relationships to the land within settler-colonial urban places, Sogorea Te' reclaimers took part in protest art, marches, and legal action. They forged solidarities with the primarily Black and Latinx food and housing justice organizations that worked in their same neighborhoods. Together they occupied urban sacred sites, like the West Berkeley Shellmound. Sogorea Te' worked within the law as well, creating a community land trust to facilitate land control transfers from local jurisdictions. In these ways, over a decade of struggle Sogorea Te' destabilized the way urban land in the Bay Area is constructed as state territory that can be bought and sold as property.

A related reclaiming unfolded a bit to the north, where the Wiyot people of Humboldt Bay (Wigi) held a vigil and symbolic occupation of the center of their territories, Tuluwat Island, for decades. White settlers had dispossessed the Wiyot of the land after a series of coordinated massacres in the 1860s. Like Sogorea Te', the Wiyot created a community land trust to regain control of the land. The Wiyot reclamation took the first step toward controlling the island when a movement organization used a grassroots funding campaign to purchase an abandoned boatyard on the island in 2000. After removing tens of thousands of tons of toxic waste from the land and replanting it in culturally and economically valuable native plants, in 2004 the city of Eureka transferred legal control of a part of Tuluwat Island to the Wiyot tribe. Movement organizations The Ongoing Return of All (Da gou rou Iouwi') and Cooperation Humboldt now work with the Wiyot's Dishgamu Land Trust to own the land in collective and manage it, decommodifying the land by making its sale as property impossible. Their efforts are now focused on ecological restoration and creating new relationships of value and production upon it.

The Ohlone and Wiyot reclaiming movements are but two of the Indigenous Pan-American reclaiming movements that take the refrain "land back" as a unifying concept. As a collection of strategies of dissent and reclaiming, Land Back movements build on centuries-long Indigenous liberation struggles, from the armed anti-colonial Indigenous resistance movements to the red-power American Indian Movement that began in the 1960s and continues its work today.

Land Back emerged out of Indigenous peoples' refusal to accept the dispossession of their homelands.[13] Grounded in practices of return and restoration, these movements work toward land back in many ways, from the state-sanctioned Ohlone and Wiyot land trusts to the fiercely repressed pipeline protest camps that continue to grow across the oil extraction landscapes of North America after Standing Rock. Indeed, disrupting state and corporate claims on places and resources within Indigenous territories has become the central way activists are building movement power. At the protest camps and across reclaimed plots, the land is a place for ideas and practices of Indigenous solidarity to unfold. As a collection of movements, Land Back represents Indigenous resurgence and reunification, a nurturing of Indigenous life and relations. Reclaiming and Land Back movements seek to decommodify the land and rebuild it as historically and socially situated territory. In doing so they seek to redefine the very meaning of land through struggle and mobilization.[14]

Overall, there is no count yet of how many places like Casiavera and Tuluwat Island there are around the world, the places engulfed by colonial and corporate dispossession and reclaimed by Indigenous peoples and rural workers. The few reclaiming stories I surveyed here are places where the usual components of states and corporations are not completely in control of the land. While many of these landscapes are still in part bound by logics of capitalist exploitation, reclaiming movements privilege small-scale, socially embedded relations to the land. Even so, these landscapes contain the memories and structures of dispossession. These are the substrate upon which reclaimers must create new forms of life.

RECLAIMERS

Scavengers, squatters, back to the landers, occupiers, radicals, peasants: the people who counter dispossession of the land and its resources and seek autonomous livelihoods go by many names. Their work gives them a particular reclaimers' knowledge and politics, as well as community.

Reclaimers take up life in capitalist ruins, but their life is not necessarily defined by them. Many reclaimers I have met did tend to have anxiety about ecological ruin, but they all also had a fundamental hope that allowed them to continue for years with the difficult task of reclaiming the land. As the lives of many in Casiavera illustrated, the allure of the city as freedom from the difficulty and drudgery of agricultural work are real. But for many, this urban allure did not hold for long. Reclaimers often returned to Casiavera after spending their young adult lives in the city as informal laborers and working in factories. These people told me that being poor in the city is a very difficult thing to be. Being a landed smallholder, however, is usually not as dangerous or draconian.

Motivated by visions of a smallholder's well-being, reclaimers turn to both direct action and legal arguments to take the land back. They act because they refuse to accept what Nick Estes has called the impossible condition of living with the state and corporate dispossession of one's homeland.[15] Instead of making do with life under such an impossible condition, movements like North America's Land Back, the Zapatistas, and Casiavera's reclaimers create new political subjectivities and agrarian possibilities to move beyond colonization and dispossession. So often locked out of legal paths forward, reclaimers in motion stop the machines, infrastructures, and commodity flows of capitalist dispossession to create new possibilities for life. In this way reclaiming is often an activity for the underground. As states and corporate elites work to criminalize reclaimers, reclaiming movements require agricultures of resistance and evasion, a kind of guerrilla gardening. This is a blessing when the state decides to ignore them and a curse when the state decides to persecute them.

Reclaimers' protests and legal machinations seek territorial control. In doing so, reclaimers reverse many of the de- and re- processes of political economy. Instead of dispossession and depeasantization, reclaimers often work toward reconnecting to the land (repeasantization). In place of deforestation, reclaimers often work toward reforestation (and by extension, deindustrialization). They also work to reverse the typical peasant political position, from reactionary smallholders resisting dispossession and social disintegration to claiming the land, agricultural diversification, and movement building. The affirmative construction of new ways of life begets autonomy. To the greatest extent possible, reclaimers live outside and beyond the state, exploring new ways of being, learning, and working that are not determined by state and capitalist domination.

Excluded from the mainstream, reclaimers rely on the resources at hand. Their labor and cooperation with their families and friends is their greatest source of productivity. Mutual aid becomes the most important kind of work. Collaboration becomes the most important kind of social relation. This is not to say individuals do not work for themselves, only that an overarching communalism is necessary. No individual can accomplish the political defense of their reclamation, or even their agricultural harvest, without others.

With meaningful salaried jobs difficult to come by and the numbers of Indonesians ejected out of agrarian spaces and into the slums and underground economy growing into the millions, getting by in rural Sumatra is as challenging as ever. The urban livelihood options open to the typical rural resident—plywood factories, motorcycle taxi drivers, security guards, domestic workers, petty traders—are increasingly hard to obtain and not all that attractive. For these reasons, reclaiming movements are not defined by their ability to bring about new land occupations, but instead by their ability to sustain reclaimers' livelihoods once they have taken the land back from the state and agribusiness. The agrarian movements and organizations are as useful as their contributions to reclaimers' diverse modes of getting by.

In large part, efforts of the landless to become smallholders depend upon bricolage livelihoods, the extreme diversification of work across space and workers' lifetimes. Repeasantization takes many paths and expressions. Nearly all of the millions of smallholder farms that have taken form in the last decade or two, from West Europe to China and the Zapatistas and Landless Workers' Movement, include forms of agroecology coupled with mobile, waged work. These new peasants have embraced livelihood bricolage as a source of creative power.

The only certainty about these reclaimers' livelihoods is that they will be diversified and polyvalent. Work for billions now means a bricolage of many activities and incomes. This is not rote labor in factories or mechanized fields. For many, industrial work no longer charts the future. Bricolage has become so ubiquitous that defining rural workers as a class distinct from urban people (or smallholders as distinct from urban laborers) is an impossibility. Rather than a marker of the transition of agrarian people away from the land, livelihood diversification is a long-term strategy of people working to stay connected to it. Put another way, turning to non-farm work is a strategy smallholders use to create and maintain their rural homesteads. Looking out into the global economy of logged-over forests, impoverished mill towns, broken-down plantations, and ruined mines gives credence to this idea.[16] Without doubt, Casiavera's reclaiming movement shows that these smallholder livelihoods can sustain a new generation of autonomous, skilled, healthy, and meaningful lives.

ONE WAY FORWARD

The state handover of public lands to corporations for exploitation of natural resources has been going on for so long, few people stop to think about how odd this practice is. Yet my conclusions after spending time with Casiavera's reclaimers echo those of agrarian justice movements that for generations have consistently, unwaveringly asked: Why not give the land back to the people? Indigenous peoples, smallholders, campaigners, politicians, and academics exist that reject the role of dispossession in our future. In doing so, they contribute to a critique of the state and corporate agrarian development as old as the Sumatran colonial plantation monocultures themselves. Amid a countryside often only seen for its corporate profits, ecological destruction, and land conflicts, land back movements cannot be written out of the histories of agrarian change. More than a few places of reclaiming have taken hold, enacting changes that speak directly to agrarian movements and the increasing need for ecological revitalization of Earth's landscapes.

Reclaiming is all the more needed as the violence and erasure of dispossession continues. Like all movements, Indonesia's agrarian mobilizations ebb and flow. On the whole, the first two decades of the twenty-first century have been a time of attenuated agrarian movement power in Indonesia. The ability to bring about a total reclaiming of Indigenous and smallholder lifeways seems far off, all the more so because new land concessions, deforestation, forest fires, labor abuses, and the severing of untold numbers of non-human relations continue apace. These dispossessions include chainsaw men (yes, they really are all men), bulldozers, private and government security forces, and thugs, all of whom are involved in a massive extinguishing of non-human life, and, in the worst cases, murders and repression of people who dare to resist too boldly.

The recent involvement of the Ministry of Defense in land grabs in Indonesian Borneo and West Papua raises the specter of still more dispossession of Indigenous peoples and smallholders at the point of a gun, this time for massive "food estates" of rice and cassava, justified as the only way forward for Indonesia to achieve "food security." That Ministry of Defense leadership created an agribusiness to serve as the corporate organ of these plantation operations alludes to these military men's desires to control both Indonesia's territories and corporate profits, all at the same time. Elites that keep one foot in professional violence and the other in land exploitation are nothing new. These regimes of agrarian dispossession transmute and change over time, but they are nothing if persistent.[17] Even as these regimes create deep structural political ecological crises, they are not likely to collapse under their own contradictions. If new reclaiming movements unfold at all like the one in Casiavera, dispossession will be countered in a piecemeal fashion over decades, as groups of activists, smallholders, and allied state functionaries mobilize to create new forms of land control and smallholder economy.

––––––––––

Back in California, I think about the volcano often. I can see in my mind's eye the paths through the agroforests, the banana gardens, and the cloud forest above it all, covering the highest reaches of the eroded caldera. Lufti planting more trees with the belief they will fund his children's education. Wangga harvesting avocados as an expression of the strength and flexibility of the Minangkabau matriarchate's unbroken control of land. Aren meeting with other activists to discuss new reclaiming mobilizations.

All of them keeping hope alive for themselves and their families, but also for all of us. Casiavera's landscape is a busy, smallholder landscape alive with an agroecology that promises certain kinds of advantages for agrarian peoples in diet, work, and autonomy, for agriculture touches on every facet of the human experience. Casiavera's reclaimers make crystal clear that any critique of dispossession is incomplete without elaboration of emancipatory ways of relating to the land and the many forms of life that can be found upon it. It is for these reasons that retellings and fragments of the history of Casiavera's reclaiming have already traveled the archipelago and across its waters on to different reclaiming movements unfolding in other nations. Still, from Canada to South Africa, rural workers worldwide need more histories like Casiavera's to inform their own efforts of building critique into action, to counter capitalist exploitation as one way forward.

During the day on the volcano, the land is full of rural peoples creating many kinds of livings. With the sun overhead and people bustling about, it is difficult to imagine how it ever could have been different. At night, it is easier to think back to the long history of forced labor, war, pogroms, and authoritarian dispossession there. Yet the final thought that comes to me about this difficult history is that tragedy, repeatedly, was overcome. Dispossession, more than once, was overturned. Reclaimers in Casiavera went beyond corporate control and ecological degradation to point to a more livable future. For those land-for-the-tiller movement members across the world who are still entrenched in difficult, long, and often heartbreaking agrarian struggles, Casiavera's history can help them construct better futures of their own.

In the end, it matters little if some remain convinced that smallholder agroecology is a "traditional" form of development and so is also archaic and irrelevant to the challenges of the Earth today. It is not so important either that others, like me, think of reclaiming movements and their agroecological landscapes as a twenty-first-century marvel. No, more important is the idea that while agroecology may not be a telos, it is a niche, a wonderful one at that, and it has already given a group of people in Casiavera a path to emancipation.

All of this is reason to appreciate what Casiavera's reclaimers have done, but it does not allow me to say that I understand their reclaiming as a phenomenon that can be repeated in other places, to similar effect. Doing so would be the same logical error as the one that underpins dispossession: that social and ecological relationships can be siloed into a single blueprint of development that can be cut out of whole cloth again and again. I think it is more likely that such a broad, comprehensive understanding of social well-being and ecological process is beyond human intelligence. For we are born of these processes of nature, not epiphenomenal to them. Even so, the deep complexities of social and ecological change should not prevent us from looking for other reclaiming movements in locales beyond Casiavera, however similar or different these other movements may be. For land connects humans to each other and other forms of life like nothing else. One of the few certainties of humanity is that land will remain at the center of any desire to create a more livable Earth, one that allows for those who work the land to control it, healthy soils, the cooling of the planet, and the survival of many forms of life, including the human.

ACKNOWLEDGMENTS

Without the chance to build with the organizers I call Malin and Aren, I could not have written this book. To them and everyone in this book who shared your story with me, I thank you. I am grateful to the person I call Tia, her husband, and her daughters for taking me into their West Sumatra home. And to all of Casiavera's reclaimers, I hope this book uplifts your experiences and movements.

The seeds of this project were planted when I was a young environmental justice researcher-activist in the California Bay Area. In those days, I worked with two mentors, Leila Salazar-Lopez and Lafcadio Cortesi (1961–2022), who showed me how to live an engaged, critical, and joyous life as part of agrarian and environmental struggles from California to Amazonia and Indonesia.

Just as it appeared the ship would never set sail for this project, William Durham ensured I would begin this research under his wise and very kind mentorship. I am eternally grateful. James Ferguson's intellectual engagement and commitment to ethnography were fundamental to this work. Zephyr Frank encouraged me to go beyond critique to celebrate the contributions of Sumatra's reclaiming movement. Shelly Coughlan and Ellen Christensen provided me calm harbors.

In Indonesia, I have been fortunate to work with Afrizal, one of Sumatra's leading sociologists. His guidance and sponsorship at his home institution, Universitas Andalas, meant everything to this project. Movement scholar J. J Polong deserves my deepest gratitude. I want to thank Virtuous Setyaka and Zeni Eka Putri, also of Universitas Andalas, for their insight during early stages of this book. Diqi and Bajau shared the gift of conviviality during fascinating and at times heartbreaking visits to the uplands. Sociologists Casiavera Nasution, Hengki Purnomo, and Wira worked with me on the agroecological survey. Fakri and Wangga Oktavery aided me in the West Sumatra archives. The staff of the American Institute for Indonesian Studies in Jakarta provided key guidance. I thank the Indonesian research agency, RISTEK, for its sponsorship.

I wrote a draft of this book while a fellow in residence at the Stanford Humanities Center. I want to thank the people associated with SHC for their support, especially Patricia Terrazas, Susan Sebbard, Andres Le Roux, and Kelda Jamison. Thanks to former director Caroline Winterer for bringing together a wonderful cohort. And for the brilliant discussions of that year, my thanks go to Asad Ahmed, Ian Beacock, Giovanna Ceserani, Henry Cowles, Charlotte Fonrobert, Alanna Hickey, Philippa Levine, Kristin Mann, Benjamin Morgan, Londa Scheibinger, Miranda Spieler, Kyla Schuller, Kate van Orden, Colin Webster, and Renren Yang.

Two years as a S. V. Ciriacy-Wantrup Fellow in Political Economy at the University of California, Berkeley, gifted me the time to write and publish. My thanks to Nancy Peluso for her generous guidance and creative, bold scholarship. I appreciate Ruxin Liu for the support during this time. Thanks to my colleagues in the Center for Southeast Asia Studies, Division of Society and Environment, Department of Environmental Science, Policy, and Management and Berkeley Geography, especially Michael Polson, Stephanie Postar, and Laura Dev for their thoughtful comments.

Rebecca Elmhirst and Laksmi Savitri deserve my special thanks. With detailed comments, these two theorists of social movements and feminisms extraordinaire sharpened my decolonial and matriarchal thinking. The chance to discuss this work at Yale's Council on Southeast Studies was invaluable. Michael Dove and Jim Scott provided both inspiration and crystal-clear ideas of how to bring together historical ecology, environmental studies, and anarchist analysis.

My thanks for the discussion and comments from a brilliant group of scholars who shared their time with me as I wrote: Tania Li, Jun Borras, Mythri Jegathesan, Sophie Sapp Moore, Kai Bosworth, and Charmaine Chua.

Maron Greenleaf, Jess Auerbach, Patrick Gallagher, Emily Beggs, Ciara Wirth, Samil Can, Nethra Samarawickrema, Johanna Markkula, Young Su Park, and Angela Garcia influenced my thinking and writing as my work on Sumatra germinated. Thank you all for the ideas, ethnographic expressions, and friendship. Kaka, Erik Wakker, Adri Zakaria, Elizabeth Forwand, and Ardith Betts, all keen observers of Sumatra, were game for many long conversations about Indonesia's social and environmental changes over the years, from which I learned much.

Additional funding came from the Wenner-Gren Foundation, National Science Foundation, Social Science Research Council, and American Institute of Indonesian Studies. My thanks to the *Journal of Peasant Studies* and *Antipode* for allowing me to rewrite and publish work that first appeared in their pages. I appreciate the attention of my colleagues, students, and friends who listened to me present work in progress during seminars and conferences from 2018 to 2022.

I thank UC Press editors Naja Pulliam Collins, Chloe Layman, and Stacy Eisenstark for their unwavering support. Two anonymous reviewers shared their deep expertise in agrarian studies of Indonesia with me, moving my analysis forward in meaningful ways.

My family gave me the love, time, and space needed for the research and writing of this book. For that, I am very grateful.

Despite my best efforts, errors remain. I take all responsibility for them.

History of the Collective Land, 1997

The following is my translation of the closing paragraph of the Casiavera Community Council document titled *History of the Collective Land*, dated February 1997.

There, please see that our children and nephews live at a level that is below average. Really, we need that land and other arable land. Our population has doubled since the first economic development plan (*Pelita*). We have almost no fields to graze our livestock. Even our terraced rice fields have been marked out for feeding our livestock. The acreage of the ex-company is empty, only having shrubs. The cultivation there has been done as a second thought. The company's enthusiasm was only at the beginning. For 28 years the company was entrenched (*bercokol*) as an investor and for not more than 10 of those years was the land tilled. Even then it was never more than one third of the land they claimed. The rest remained as a breeding place for forest pigs that devour (*memakan habis*) our people's crops. This 'seasonal investor' would only do anything if there was another joint investor who would flirt with them (*dirayukan*), otherwise bushes and forest pigs are its most prominent investment. Is such an investor proper (*pantaskah*), who defeats the aspirations of our people who live below the average, who need to raise their standard of living, who have for so long desired an equitable distribution (*pemerataan*)? We as the council request (*memohon*) and demand (*menuntut*) the return of our collective land, we are not just requesting and demanding now, we have demanded this from the start. We have created a program to make the collective land a pilot project to eradicate (*mengentaskan*) poverty from the countryside (*korong kampung*) thoroughly and finally. The program has been created by our nephews who have migrated out of the community to the city, they are the ones who will provide the capital and the technology required for that. Every time we watch TV, we are very impressed with the determination and sincerity of President Suharto, who is so enthusiastic in striving to fight poverty in our homeland, so that a just and prosperous society based on *Pancasila* and the 1945 constitution will come about at the start of the twenty-first century. It is our heart and conscience as the council that we do not want our children and nephews to be left behind and pitied anymore in the 21st century. We want to be a pioneer (*pelopor*) in the development of our village here on the slopes of the Aren volcano. At the time of the revolution and the Dutch aggression all through the year 1948/49 our community became famous as a basis for the struggle for achieving freedom. And now there is an opportunity for our children and nephews to pioneer the development for which as our dear father President Suharto has beaten the gong (*dideggungkan*). We do not just hope for opportunities, we need them, and the collective land is a golden opportunity for our people.

Indonesian Peasant Union (Serikat Petani Indonesia) Charter Documents, 1998

DECLARATION OF FORMATION OF THE INDONESIAN PEASANT UNION FEDERATION (FEDERASI SERIKAT PETANI INDONESIA)

Indonesia is an agrarian nation whose economic development should be based on sound agricultural development on which the development of the state and its people as a whole should be founded. In reality, the state has implemented a form of development which is not in the interests of or in accordance with the characteristics of the Indonesian people.

We, the peasants of Indonesia have, in an organized and planned way, made every effort to create development practices based on the needs of the people and agricultural conditions of the state. Our struggle has taken shape in the form of peasants' organizations and various actions. We the peasants of Indonesia have not ceased in our struggle, although this struggle did not receive the support of the political system of the New Order regime. In order to attain a peaceful and beneficial existence for the peasants and the people of Indonesia via agricultural reform and development,

We denounce the authoritarian political system which has chained the peasants and denied them their rights to freedom of expression and association. The authoritarian political system has violated the peasants' rights.

We reject the capitalistic economic system which has resulted in the theft of peasants' lands, destroyed the environment and forced the peasants into a biased and unfair trading system, as well as violating the rights of traditional peoples.

We reject a cultural system which is not egalitarian and emancipatory. Besides the violation of the peasants' political and economic rights, the peasants' cultural systems have also been eroded. Peasants have been isolated, left behind in terms of economic development, experienced social inequality, including inequality in the relationship between gender. A culture based on consumerism, individualism and inequality has destroyed their social, moral and ethical foundation.

In order to overcome all the problems faced by the peasants of Indonesia, with God's Mercy, we, the peasants of Indonesia, found the FSPI in order to reclaim our right to self-determination and to struggle for a just agrarian society and democratic political system.

Dolok Maraja, Asahan, North Sumatra, Indonesia
23.55 WIB
8 July 1998

Signed:

1. Representative of Organisasi Petani Aceh Aceh
2. Representative of Serikat Petani Sumatera Utara (SPSU)
3. Representative of Badan Perjuangan Rakyat Penunggu Indonesia (BPRPI)
4. Representative of Organisasi Petani Sumatera Barat
5. Representative of Organisasi Petani Riau
6. Representative of Organisasi Petani Jambi
7. Representative of Organisasi Petani Bengkulu
8. Representative of Organisasi Petani Sumatera Selatan
9. Representative of Persatuan Insan Tani Lampung (PITL)

1ST DECLARATION OF THE INDONESIAN PEASANT UNION FEDERATION

On 8 July 1998, in Kampung Dolok Maraja, Bandar Pulau, Asahan, North Sumatra, representatives from 9 peasant organizations announced the formation of the Federation of Indonesian Peasants Union (FSPI).

Background:

1. The authoritarian rule of the New Order regime did not allow for freedom of political expression nor the freedom of association and participation, especially by the people.

2. The New Order regime established a capitalist economic system which encouraged the excessive exploitation of natural resources and the people of Indonesia by monopolies in the hands of a small group and class. Throughout the 32 years of the New Order regime, land has been taken away from the people and turned into industrial areas, golf courses and capital-intensive agribusinesses.

3. The government deceived the people with its claims of economic success, as is proven by the current economic crisis. Economic development has been pushed right back to beyond 'point zero'.

4. Despite the severe economic crisis, the New Order regime still attempted to rule with its existing authoritarian political format, notwithstanding that it claimed legitimacy for its rule on the basis of economic development. That this political system is redundant is evidenced not least by the riots and unrest throughout April and May of this year.

5. The economic crisis resulted in the demands for reform and the fall of the New Order regime. Nevertheless, reform cannot be attained, until Indonesia's economy is based on the development of the agricultural and farming sector. The development of these sectors includes both economic and political development.

In accordance with the above, the foundation of FSPI is a manifestation of the commitment and consistency of the struggle by the peasants in Indonesia in general and represents the continuation of the struggle of the peasants of Indonesia throughout this period. FSPI struggle for reform is based on Religious Belief, Solidarity, Independence, Democracy, Non-Violence and Non-Discrimination. FSPI's commitment to this struggle will take the form of a number of programs concerned with Agrarian and Social and Political Reform.

FSPI proposes to:

1. Lobby the government to implement Land Reform policies and to recognize traditional land rights.

2. Promote and implement organic farming practices and the sustainable exploitation of natural resources.

3. Develop farming cooperatives.

4. Promote freedom of association and to provide a political role to the peasants of Indonesia.

The first concrete step taken by FSPI to implement its principles is to assist the peasants of Lampung whose harvest has failed, by calling for the support and solidarity of all Indonesian peasants, particularly in Sumatra, and to send them rice and other produce.

Medan 12 July 1998
FEDERASI SERIKAT PETANI INDONESIA (FSPI)

Counter-Mapping

To illustrate the changes that unfolded across Casiavera's landscape in fine detail, my study made use of technologies of mapping and satellite photography. Powerful settler-colonial elites developed and used both maps and satellite photography as technologies of control of space and territory. To not replicate the domination so often enacted through spatial technologies of surveillance and mapmaking, I developed a reflexive and collaborative approach to my maps as an effort at counter-mapping.

Taking seriously my desire to be a collaborator, ally, and accomplice to Casiavera's continued efforts of countering dispossession and reclaiming, I first acknowledged that despite the uptake of remote sensing and geographic information systems (GIS) technologies by smallholders and activists to aid in their efforts at social change, there was a complete absence of maps of any kind in Casiavera created by stallholders or the family-lineage leaders in the archives, offices, or in people's homes. Casiavera's reclaimers and other residents did not engage in any counter-mapping of their own; there was no mapping practice in Casiavera leading up to their reclaiming, and one has not been developed after it. No one who lives in Casiavera has illustrated their own counter-maps to represent the importance of their land to them, nor did I meet anyone who has access to the internet, computers, handheld GIS units, and software required to create digital counter-maps.

Into this absence I brought my own spatial information. I made maps that brought my own perspective and will into a domain of knowledge where objectivity is especially elusive, starting first with the decision of what to map. I chose to map vegetation cover and human infrastructure, because they were the most visible expressions of the changes that have unfolded across Casiavera. I created these categories with the principles of counter-mapping in mind. I drew only on publicly available data, I used social theory and qualitative research, and I made them while in conversation with the people who have lived the social and environmental changes I sought to map.

I then brought my maps to Casiavera. I did so to build upon reclaimers' perspectives and ideas about them. Reclaimers without fail identified their own plots in the images and followed the landscape depicted in the images from their own plots to their neighbors' plots and the edges of the customary land. Even as they identified their own and their neighbors' plots, told detailed remembrances of the land through time, and often remarked on the species of trees and grass that were shown growing in the maps, what I found most remarkable was what reclaimers pointed out to me to be missing from my maps. I presented the most detailed of all satellite photography available to the public (of course classified spatial data can be assumed to be more advanced than the public WorldView data I used). But people often remarked that the boundaries of their own plots and Casiavera's customary land were not properly depicted or were missing altogether. Where on the ground, smallholders see these borders to be evident, in the satellite images they were "difficult to see," "not clear," and "gray." They said the same about the footpaths and motorcycle tracks that connect their homes to their plots. When traveling the customary land, these paths also contribute to the clearly defined perception of the landscape being a mosaic of individual plots.

That the reclaimers were concerned that the maps obscured the boundaries of their plots spoke directly to the political effects of spatial information and technologies: even the most highly detailed satellite photographs obscure social relations to the land. Although my maps made crystal clear individual trees on the land, they did not include depiction of the boundary markers and footpaths that exist obscured under the trees' canopy. Neither did they include any of the social cues of land delineation that are easily detectable during a walk through it. Casiavera's forests were cultivated from the ground, and it is only from the ground that their full sociality can be read. Flying overhead, satellites remain unable to represent the land in a way that can trace how it is perceived and used by those who live and work on it.

My maps seek to illustrate the dynamism of Casiavera's reclaimers' use of the land, but most often maps do not. Counter-mapping must be diachronic, otherwise it risks replicating maps of the state and corporations that are a projection of the desire to remake space into state-centric forms. In Indonesia, none of the maps produced by the Dutch Forest Service or the Republic's Ministries of Forestry and Agriculture, or the modern National Land Agency and agribusiness companies, are diachronic. All these maps depict only a single time period, what these maps give to be an amorphous and ongoing "present."

Take two modern examples of state maps relevant to Casiavera. Both depict a single time period, the "present," and provide only categories of information that support the Indonesian state's ongoing claim to control Casiavera's collective land. The National Land Agency made the first map in 2011.[1] It was published as part of a Land Agency survey carried out in response to the Casiavera Community Council's allegation that Dona Company, holder of the industrial use permit in Casiavera, had abandoned the land. The map is part of a document that reported the results of a Land Agency survey team's visit to Casiavera and reads, in part, "the land is controlled by the people," and "the land is not controlled by the rights holder, Dona Company." Despite the report's finding of Casiavera reclaimers' de facto control of the land, the map included in the report reproduces many of the problems with state and corporate mapping technologies. Carrying the National Land Agency insignia, the map depicts the boundaries of the old company land concession and categorizes agricultural systems within the land concession boundaries. It depicts the land as being used completely for "mixed gardens" (*kebun campuran*), which is one of the many terms the government uses for agroforests. It gives additional limited specifics of the cultivars growing in these mixed gardens (cinnamon, banana, coffee, betel nut, and tobacco). It also includes representation of two footpaths (*jalan setapak*) and several homes (*rumah*). In the legend are a few items that do not appear in the map itself (border marker point, road, and district boundary).

In its form, the map follows the convention of Western mapmaking. It is rectangular, depicts the cardinal directions accurately with north as up, uses a scale of 1:10,000, and uses a simple black ink illustration method to clearly depict boundaries and labels. Striking is the lack of any context to the map, without any indications of topography, land use, or infrastructure outside of the concession boundary. What's more, the map uses a map projection system to assign the coordinates of the company's land concession boundaries that I have been unable to decipher or convert to the common Universal Transverse Mercator coordinates system in use today, making the map impossible to locate in space in Casiavera. Finally, the existence of individuals' plots is conspicuously absent on the Land Agency map. The map instead labels five undefined regions within the concession as mixed forest, creating a visual representation of a space that is undifferentiated or perhaps cultivated by a few groups of people. And so, while the National Land Agency map provides an authoritative perspective on the reality that reclaimers used the land in 2011, it does so in a weak way. The map does not give any useful information about the decades-long cultivation of the land.

The second masters' map, the National Land Agency Online Map, also erases the presence of the reclaimers. Like the 2011 map, this map is a projection of state will and authority in the "present"; it carries a production date of 2017, although I know this map to have been created at least as early as 2015.[2] And so this map makes a persistent claim to represent the present. The Online Map is more sophisticated than the 2011 survey map and is viewed through an interactive web browser interface that allows the viewer to change the location and scale of the map. In many ways, this map contains definitive conventions of modern Western mapmaking. It is interactive. It can change scales. It presents north as up and includes an on-the-fly rendering of its scale listed in both meters and feet, and it has clickable "layers" that display spatial information. The base map of rivers, roads, and administrative boundaries is a well-known and widely used GIS, called Open Street Map. In addition to the Open Street Map base, the Online Map presents a single category of information: ownership of land plots. These are displayed in eight categories, including among others private property (*hak milik*) and state land-lease concessions.

Panning and zooming to show Casiavera reveals that in this map, the National Land Agency fails to depict any private property land plots existing in the region.

The only land plot in Casiavera shown in this map is a green-shaded polygon, labeled as a land concession. Its precise geographic coordinates cannot be determined from the web browser interface, but its general location and shape suggest this is the Dona Company concession. The map is a projection of the National Land Agency's position that the Dona Company exists, despite having no presence at all in the real world that the map claims to represent.

The Online Map is a pure spatial visualization of state desire and will, nothing less and nothing more. What's more, the National Land Agency recognizes that its authority to project its spatial desires requires suppressing other people's efforts to use maps. In order to access this map, I had to sign an online user agreement by clicking a box that reads "I agree" to the following terms: "The maps and information contained in this page should not be disseminated in other formats, either as digital files or prints. Access to this map is only allowed from www.map.bpn.go.id." The Land Agency chose to create its Online Map to illustrate its view of legal use of the land and made it publicly available on the internet to viewers. Far from following egalitarian ideas of open access, in order to preserve its authority to influence how land is used, the agency forces its viewers to promise to never share this information among themselves or use this information to create counter-maps of their own. Such is the dependence on the form and presentation of maps; the Land Agency seeks to limit viewers' visualization of land plots to the digital environment of the Land Agency online web browser.

In this case, counter-mapping does not require making new maps of my own. It would only require that I share this Land Agency map with others. I am unclear what the subversion of the National Land Agency's rules about viewing and sharing its map would accomplish, but often acts of subversion do not have certain effects. The only certainty of subversive resistance is that it can create new, undetermined possibilities for action.

Alas, it seems likely that the Online Map does not have much to contribute to the reclaiming movement in Casiavera. Reclaimers there know all too well that the National Land Agency never canceled Dona Company's land concession on the collective land. Most know they have only de facto control of the land. Most also know that many politicians going all the way up to their governor have stated their support for the Casiavera reclaimers' position. Still, the Dona Company could one day return if it could muster sufficient state support.

I made counter-maps in dialogue with reclaimers in Casiavera, and for that reason alone I was able to avoid many of the mistakes of the National Land Agency even though I, too, used the conventions of Western mapping. For one thing, instead of making maps of the region, as a state planner would, I mapped the local landscape. Too often, as with the 2011 survey map, larger-scale regional maps obscure social realities as homogenous units to erase and obscure the actions of smallholders and their communities. Maps are important not only for what they include, but also for what they leave out.[3]

I also avoided the mapping of administrative boundaries, because governments, nongovernmental organizations, and corporations use maps to classify, register, and monitor territories that condone or make illegal certain forms of land use. These kinds of maps are typically tools of state power, often used for controlling populations.[4] Why map administrative boundaries if they are neither certain nor important for the work, as I hope my maps will be? Finally, and most importantly, my maps all show changes in time. And it is these diachronic visualizations that provide the real insight into the place.

Only with deep context, real-world engagements, and appropriate conceptualization of scale will satellite photographs and other kinds of mapping break free from their role in aiding diverse ecologies to be targeted for destruction and erasure and the consolidation of exploitative political forests that deny local land users' access. Such counter-mapping is not always required to reclaim land. But it can be useful for these land struggles, because satellites' unique gaze provides representations of land reclaiming as it ages and matures, spurring new, visually informed ideas.

NOTES

INTRODUCTION: LAND BACK

1. I attended the summit at the invitation of a friend who worked for the Indonesian Forum for the Environment. The Via Campesina declaration was published in Indonesian by the Indonesian Peasant Union: *Deklarasi Bukit Tinggi: Reforma Agraria dan Mempertahankan Lahan dan Wilayah di Abad ke-21* [Bukit Tinggi Declaration: Agrarian reform and maintaining land and territory in the 21st century], 13 July 2012. I have a copy of the declaration.

2. Because of the fraught situation in the Sumatran countryside, I use pseudonyms for most people and places in this study.

3. *Tanah ulayat* is a common term in Indonesia that I choose to translate as collective land. The term could also be translated as customary land, traditional land, or indigenous land. The key idea of the term is that it is a plot of land that is owned not by an individual but by a defined community as a collective.

4. Indonesia's agrarian movements are vast and diverse, making a definitive count of reclaimers and their lands impossible. The Indonesian Peasant Union says it reclaimed 200,000 hectares of land from 2007 to 2013. The Consortium for Agrarian Reform (Konsorsium Pembaruan Agraria) in Bandung, Java, has since 1994 kept a running list of more than four thousand land conflict sites involving hundreds of thousands of hectares of land slated for plantations, logging, mining, and other state and agribusiness exploitation. Additional sources suggest this is only a partial accounting. Sawit Watch, in Bogor, Java, documented nearly three thousand reclaiming movements at industrial palm oil plantations since 1998. Analysis by Sumatran agrarian justice organizations ScaleUp and CAPPA show that from 2007 to 2011 reclaiming movements in five provinces of Sumatra unfolded at state land concession leases totaling 1,500,000 hectares of land and included tens of thousands of rural peoples.

5. Analysis and estimates of forest death are another contentious arena of knowledge production. The accuracy and validity of deforestation rate estimates vary widely. Two of the most rigorous estimates of Sumatran forest change over time indicate the nearly complete loss of Sumatra's tropical lowland rainforests and widespread degradation and fragmentation of the extant hill and montane rainforests by 2010 (Margono et al. 2012; Brioch et al. 2011).

6. The Intergovernmental Panel on Climate Change (2013) provides estimates of the sources of climate-changing gases.

7. The agrarian question is one of the fundamental, unifying threads of social inquiry, taken up again and again by scholars interested in understanding how our world is changing. Compare Kautsky ([1899] 1988), Chayanov ([1922] 1986), Watts (1983), and Bernstein (2010).

8. My understanding of dispossession emerges from my time with Sumatran smallholders and agrarian activists and recent texts from Nick Estes (2019) and Audra Simpson (2014), especially.

9. Robert Nichols's (2020) critical theorization of dispossession provides a bringing together of ideas of dispossession across diverse domains of scholarship, time periods, and social movements.

10. Touchstone studies that detail the terrible costs of dispossession for industrial agriculture include Durham (1979), Stoler (1985), Guthman (2017), and Li and Semedi (2021).

11. Lenin elaborated his antipathy toward smallholding in the June 1920 pamphlet *"Left-Wing" Communism: An Infantile Disorder.* As Lenin rose to be leader of Russia's ruling Communist Party, anti-peasant political thought became official state policy. The concept of depeasantization led Lenin's elite politburo to direct the large-scale forced collectivization of the countryside (see Tarrow 1998, 12). Under Stalin, the forced collectivization scheme would eventually fail under a specter of violence and starvation, suggesting the depth of peasant resistance to these policies (Netting 1993, 23).

12. One illustration of how smart people refuse to let go of the idea that smallholders are a hindrance to modernity is Cargill CEO Greg Page's February 10, 2011, presentation to Stanford University's Center on Food Security and Environment, "Balancing the Race to Caloric Sufficiency," in which he spoke about the existence of smallholders as a costly "rural sociology premium" that the global food system would be better off without.

13. The white paper "A Plausible Vision to Feed the Planet" from the Breakthrough Institute (2018), an increasingly influential ecomodernist think tank, provides insight into the damning way the teleology of dispossession functions. Without presenting any evidence, the institute writes that the historical movement of smallholders away from the countryside toward urban centers "has greatly improved people's lives." Building on this assumption of progress, the institute argues that because powerful "political and economic developments" exist that underpin a "robust, ongoing global trend" of the expansion of industrial agriculture, smallholder agriculture must be unfeasible.

14. Some agrarian scholars have been too taken with a Leninist-influenced political economy that tended toward an aversion to the peasantry. These include Eric Hobsbawm, who in his 1959 treatise on peasant movements concluded that they would be left behind at the threshold of modernity.

15. This lineage of critical scholars owes much to Alexander Chayanov for demonstrating how communal land control and smallholdings have inherent advantages over capitalized and input-dependent industrial agricultures. A politically engaged agricultural scientist, Chayanov lived just long enough to see the rise of Lenin's anti-peasant politics and the destruction of the Russian agricultural cooperatives Chayanov had worked to establish before he was summarily executed, labeled a dissenter for his anachronistic agrarianism that placed the smallholder as a permanent, productive component of society.

16. Indonesia's 2018 Inter-Census Agricultural Survey (Hasil Survei Pertanian Antar Sensus) estimates the number of Indonesian agricultural workers to be some thirty-three million, living in twenty-eight million households that include ninety-eight million people. The survey also estimates that the total number of both landed and landless agricultural households had increased since 2013 (2018, 24–26, 49).

17. James Ferguson (1999) offers the metaphor of a bush to think about social development beyond linear progressions through time. Charles Darwin (1837), often incorrectly associated with the origination of linear assumptions of social change, offered that development was best envisioned as the growth of a coral reef, with each generation of life growing outward on the solidified substrate of previous generations.

18. McMichael (1997).

19. By the early 2000s there were millions of landless workers and smallholder farmers reclaiming the land across the global countryside. As Sam Moyo and Pari Yeros (2005) wrote, these reclaimers were nothing less than the resurgence of rural movements across Africa, Asia, and Latin America. Diverse and differentiated in many ways, these "new social movements" organized around struggles of subaltern identity, autonomy, and ecological crises.

20. The quote is from Marcos's (2003) essay "Another Calendar: That of Resistance." Marcos began writing about smallholder mobilization as working toward "something new" in 1992, in the influential essay "The Southeast in Two Winds—A Storm and a Prophecy."

21. Of these, works by Jan Douwe van der Ploeg (2012, 2013), J. Martinez-Alier (2011, 2002), and Miguel Altieri and Peter Rosset (1996) deserve mention for the way they have foregrounded the continued importance of a smallholder, modern-day peasant agriculture to both economies and environmental movements.

22. My discussion of degrowth here draws on Claudio Cattaneo and Marc Gavaldà's (2010) linking of degrowth, land-reclaiming movements in the hill forests around Barcelona, and anarchist organizing. That Kohei Saito's (2022) elaboration of a Marxist degrowth communism has sold over a half million copies in a manner of months speaks to the power of post-development theory. In response to the

emergence of post-growth ideas, Alexander Dunlap (2020) has asked if "degrowth can struggle," remarking on the fact that scholars of degrowth have in general been too removed from the agrarian and environmental movements that elaborate many degrowth ideals.

23. The revolutionary collectivist Michael Bakunin ([1873] 1971) described the power of anarchist direct-action protest across Europe a century and a half ago. David Graeber (2004), a movement intellectual, found power in the North from the non-violent, direct-action repertoires that took form and grew in the twentieth century. Oceti Sakowin Oyate scholar Nick Estes (2019) elaborated Indigenous lineages of direct action and the way they created newfound power confronting geno-cide, removal, and dispossession.

24. Arturo Escobar (2018) writes about how resurgent movements are making new worlds of autonomy, interdependence, and collectivity. Within this way of under-standing reality, well-being for all (conviviality, *bien vivir, sumak kawsay*) requires cre-ation and affirmation of many diverse worlds within the world, the pluriverse.

25. Graeber (2009, 437).

26. Formative works in this tradition include Anthony Winson's (1982) analysis of a single Prussian plantation estate that rewrote Lenin's analysis of agrarian change. Philippe Bourgois's (1988) examination of race and livelihood in a single United Fruit Company plantation also comes to mind.

27. Evans-Pritchard (1950) most clearly laid out the need for anthropology to over-come its World War II–era tendency toward anti- and apolitical analysis in the 1950 Marett Lecture at Oxford University.

28. In a sophisticated discussion, Bourgois and Schonberg (2009, 15) remind us that photo-ethnography is in no way unproblematic. Still, when done ethically, using photographs as part of ethnographic practice can bring aesthetics and politics into social theory in meaningful ways. I join a growing group of ethnographers who build on Bourgois and Schonberg's work to pair photography with contextual and theoreti-cal analysis (e.g., Jason de Leon and Kregg Hetherington). As part of this method, ethnographers may or may not use pseudonyms and may choose to reveal some faces in publications.

1. UNDER THE GUN

1. Of the scholars who have worked to both understand the world and change it, traditions in critical agrarian and social movement studies have proven exceptionally informative (e.g., Bjork-James 2020; Borras 2016; Estes 2019).

2. In addition to many interviews with residents of Casiavera, I piece together this account from four documents in Wibawa's Casiavera archive: Provincial Agrarian Inspectorate, Land Use Permission Letter (*Surat Izin Pemaikaian Tanah*), 3 June 1968, state declaration addressed to the governor of West Sumatra; Casiavera Com-munity Council, Request That PT Dona's Application for Permission to Extend Its HGU on Casiavera's Customary Land Should Be Rejected (*Mohon Agar Permintaan*

Izin Perpanjangan HGU, PT Dona atas Tanah Ulayat Nagari di Casiavera supaya ditolak), 24 December 1996, letter addressed to the governor of West Sumatra; Casiavera Community Council, Update and Request for Rejection (*Pemberitahauan dan Mohon untuk Pemblokiran*), 11 April 1997, letter addressed to the provincial head of the National Land Agency; and Casiavera Community Council, History of the Land Owned by Casiavera (*Riwayat Tanah Milik Nagari Casiavera*), February 1997. The author has a copy of Wibawa's archive.

3. Of the robust scholarship on the Gestapu and the events that brought forth the political violence and authoritarianism of the New Order, John Roosa's *Pretext for Mass Murder* (2006) provides the clearest rendering of how New Order army killers justified their actions by imagining their victims as part of the Gestapu, an imagined movement of deranged leftists. Geoffrey Robinson (2018, 166) documents how army anticommunists transformed the name of the perpetrators of the 1965 killings of the generals from the September 30th Movement to the Gestapu, what the chief of the Army Information Center at the time called a "Gestapo-like terror," with connotations of evil.

4. Geoffrey Robinson (2018, 120–21) carefully reviewed scholarship on the pogrom to conclude that the figure of five hundred thousand is a conservative estimate, with significant uncertainty stemming from the lack of any government support for investigations into the killings, the speed of the executions, and the fact that the perpetrators dumped their victims in waterways and countless unmarked graves. As Robert Cribb (2002, 558) writes about the uncertainty surrounding the numbers of victims of the pogrom: "All that is available are the educated guesses of a number of variously informed people whose judgments are based on what they think is plausible, considering the anecdotes with which they are familiar and their overall knowledge of Indonesian society and politics."

5. See the oral histories from West Sumatra in Tempo, "Requiem for a Massacre," *Tempo*, 1–7 October 2012.

6. Abril told me that the cruelest part of the accusations and threats was that he and the others had made great sacrifices to "crush" (*menumpas*) the few Communist Party members who had lived in Casiavera only years before. He claims that some of the people in the room even had family members there who were the victims of "communist barbarity" (*kebiadaban*).

7. Kuman and Vania's story came to me when I was out in their plot harvesting elephant grass with them on the collective land on October 14, 2016.

8. The victim of the beating has passed away. This story was told to me in an interview with the man's brother, Zed, on April 12, 2016 at his home in Casiavera. Zed was at the time a policeman.

9. Ironically, the HGU could only be issued for state land. And so the state engaged in all kinds of strategic positionings to define agriculturists' land as state land in order to issue a land lease (Afrizal 2007, 156).

10. The final 1968 Hak Guna Usaha document for the Dona Company has never been seen publicly, despite many efforts by people in Casiavera to see it. Staff at the

modern incarnation of the Agrarian Directorate, the National Land Agency, told me that they have the document but that it is private. However, the documents used to create the final land concession permit are public, and I hold several of them.

11. Throughout the decades of New Order rule the Agrarian Directorate would become the National Land Agency in 1988, to adjudicate "land-related issues," create spatial plans, issue land concessions, and enforce "land security" (Rachman 2011).

12. Peluso (1992, 50).

13. The first Dutch land appropriations on any significant scale were the private landed estates (*particuliere landerijen*).

14. See Michael Taussig's (2018) ethnography of the way the Colombian state constructed new powers across the landscape through the entwined biological and social horrors of oil palm plantation expansion in the lowland forests.

15. After the European states attacked smallholders' collective land control within their own borders in the nineteenth century (Netting 1993), these same states then turned their efforts to enacting rural enclosures in their colonial territories (Fox et al. 2009; Peluso 2011). Through it all peasant forests of the kind Ramachandra Guha (1990, 188) writes about in the Himalayan Chipko "represented a defense of a traditional economic and social system which afforded the peasantry stability."

16. For Robert Nichols (2020, 83, 88), colonial dispossession is a process whereby government and corporate elites apportion a grid of property across Indigenous lands, with the land concession playing a central role in the manifold techniques of dispossession that force people from the land.

17. Michael Dove (2011, 23) and Peter Vandergeest and Nancy Peluso (2006) consider the land concession to be a technology of social control that colonialists applied across Southeast Asia, with devastating effects for smallholder economies. Rebecca Hardin (2011) has made parallel conclusions about the concession in Central Africa.

18. P. J. B. De Perez, *Noordelijk gedeelte van het Gouvernement van Sumatra's Westkust volgens de nieuwste van hijdrographische opnemingen en voor de binnenlanden volgens de opnemingen en M.S.S. Kaarten*, 1842, University of Leiden, no. 05410-05.

19. A. J. Bogaerts, *Kaart van een gedeelte der Westkust van Sumatra, Breda: Bogaerts*, 1855, Leiden University Maps Archive, no. 05867.

20. J. W. H. Cordes, *Provinciale Verordening van de Preanger Regentschappen betreffende het tegengaan van boschdiefstallen en boschbeschadiginen* [Provincial regulation of the Preanger Regencies concerning the prevention of forest thefts and damage to forests], Javasche Courant, 25 February 1896, no.16, p. 2 (NI), no. 5597, in Van Goor (1982).

21. Bhandar (2018) illustrates how colonial cadastral surveys and mapping of territory made possible the dispossession of peasant and agricultural worker communities.

2. PRIMITIVE ENCLOSURES

1. Christine Dobbin (1977) presents detailed archival records of the first arrival of coffee in West Sumatra as part of her history of the origins of the Padri rebellion

(1803–37) on Aren and across the region. Gordon Wrigley (1988) provides documentation of coffee's arrival in West Java.

2. This section of my account of the forced cultivation system on Aren draws on Dobbin (1983) and Oki (1977, 23, 31, 35, 94). Elson (1994) and Fasseur (1992) provide shocking accounts of the draconian conditions of forced cultivation in Java.

3. A few larger plantation areas of coffee were also planted in Sumatra during this time, but they were not part of the forced cultivation system that all Sumatran smallholders were subject to. Most coffee planted as part of forced cultivation was in patches and hedgerows, and coffee was the first widespread form of monoculture (Dobbin 1983, 40)

4. Elizabeth Graves (1981) gives a comprehensive account of the rise of surveillance and policing as part of the colonial coffee cultivation system and Minangkabau resistance to it.

5. *Akte van Erfpacht No. 28, Verspondingnummer 3/202*, December 1905, Yayasan Lembaga Bantuan Hukum Indonesia Archive, Padang, West Sumatra. I have a photocopy of this document.

6. J. H. De Bussy, Halaban Cultuur, in *Handboek voor Cultuur en Handelsondernemingen in Nederlands-Indië*, 1925, Colonial Business Indonesia Database, University of Leiden.

7. W. G. Nieuwenhuysen, *Afschrift No 2906/11*, 18 November 1918, Yayasan Lembaga Bantuan Hukum Indonesia Archive, Padang, West Sumatra.

8. Het Hoofd, *Landmeterskennis No. 129.* 12 September 1919, 1919, Padang; and Paehlig, Griffier Meester Rudolf, *Acte van Erfpacht No. 27*, Yayasan Lembaga Bantuan Hukum Indonesia Archive, Padang, West Sumatra.

9. The Dutch spent nearly two hundred years as merchants in the archipelago before they became occupiers. They established the port of Batavia in 1620. Forty years later, in 1666, the Dutch invaded the West Sumatra port of Padang, wresting control of it from the Acehnese kingdom, which was at war with the prince of Minangkabau in the uplands. In the late 1700s the Dutch lost the port in a Minangkabau invasion, and for several decades afterward the Dutch were weakened by a ten-year naval blockade of Sumatra's West coast by the British, who had settled much of the coast at the time (Graves 1981).

10. Minangkabau society was formed around women's power as heirs, heads of households, and controllers of land (Sanday 2002; Blackwood 2000). Across centuries of colonial and capitalist attempts at domination, women's power has been fundamental to the ways the Minangkabau *nagari* have been able to reconstitute and recreate themselves as sovereign polities.

11. The formation, repression, and reconstruction of the *nagari* as unique and invaluable Muslim, matriarchal, and self-governed polities has been a central concern of anthropology, sociology, and Indonesia studies from these fields' troubled colonial origins. Exemplary, revisionary post-colonial analysis of the social and political history of the *nagari* includes work by Taufik Abdullah (1966), Franz von Benda-Beckmann and Keebet von Benda-Beckmann (2011, 2013), and Jeffrey Hadler (2008).

12. Both Kahn (1993, 70–73, 156) and Oki (1977, 261) discuss land dynamics of the *nagari* in detail.

13. Casiavera's community council meeting center holds a register of "community heads" (*kepala nagari*) from 1837. The first listed is Datuk Masaid, possibly the first to be elected under the system of having a single, male, state-recognized leader. Alongside the names and dates of rule is a column listing the major historical epochs of the leader's political period, including "Dutch Colonialism," "Japanese Invasion," "New Order," and "Reformasi." I noted this genealogy during a visit to the *nagari* offices on October 12, 2015.

14. Hadler (2008, 48, 58).

15. Later, the New Order would attempt an even more severe curtailment of the *nagari*, when in 1983 the state declared them illegal and created new, smaller Javanese-style villages (*desa*), constructed to be more easily controlled and policed. For the next two decades the *nagari* and their forms of customary rule were almost completely sidelined and co-opted by the military regime. The *nagari* were suppressed but never eradicated allowing Minangkabau customary leaders, activists, and lawyers to achieve the restoration of the *nagari* as legal state entities and the essential polities of Minangkabau life just a year or two after the fall of the New Order.

16. A. J. Beversluis, *Dienstkring Sumatra's Westkust: Maandrapport over Januari, Februari 1923*, [West Sumatra Service Section: Monthly reports January, February 1923], 1923, unpublished, pp. 3+3, app. (NI), no. 4026, in Van Goor (1982).

17. Beversluis, *Dienstkring Sumatra's Westkust*.

18. Bos (1911) and Socfin (2011).

19. C. H. Japing, *Dienstkring Sumatra's Westkust en Tapanoeli: Verslag over het jaar 1927* [Service Section Sumatra's west coast and Tapanuli: Annual report for 1927], 1928, unpublished, pp. 118, app. (NI), no. 4334, in Van Goor 1982.

20. Kahn (1984, 315–19) details the full range of industry that was the source of the dispossessions. Not to be overlooked was the deforestation needed to power these industries. The expansion of the Sumatran railroad required timber for rail sleepers, and the giant Ombilin coal mine consumed vast amounts of wood to frame mineshafts and power the mine's steam engines, starting in 1928. Pelzer (1978, 90) witnessed the upheavals of colonial dispossession in West Sumatra firsthand.

21. Beversluis, *Dienstkring Sumatra's Westkust*. On dispossession leading to union organizing in West Sumatra, see Kahn (1984, 298).

22. Dienst van het Boschwezen, *Jaarverslag 1915: Sumatra's Westkust, Tapanoeli en Atjeh* [1915 annual report: Sumatra's west coast, Tapanuli and Aceh], 1916, unpublished, pp. 26 (NI), no. 4085, in Van Goor (1982).

23. Boschwezen, *Jaarverslag 1915*.

24. This is E. P. Thompson's (1975, 31) fundamental insight in his study of peasant resistance in feudal England. The conclusion remains equally valid in the case of the Bukit Barisan.

25. J. Ballot, *Ontwerp agrarische regeling voor Sumatra's Westkust, Mei 1911*, Padang: De Volharding, Padang, Leiden University Library Collection KITLV3 M w 155.

26. H. J. Kerbert, *"De besternming te geven aande opte heffen koffiereserven (I)"* [The destination to be given to the coffee reserves which are to be abandoned (I)], *Tectona* 8 (1915): 257–82 (NI), notes on the article of T. Altona (*Tectona* 6: 296–326), no. 5460, in Van Goor (1982).

27. C. H. Japing, *Dienstkring Sumatra's Westkust en Tapanoeli: Verslag over het jaar 1928* [Service Section Sumatra's west coast and Tapanuli: Annual report for 1928], 1929, unpublished, p. III, no. 4335, in Van Goor (1982).

28. C. H. Japing, *Dienstkring Sumatra's Westkust en Tapanoeli: Verslag over het jaar 1927* [Service Section Sumatra's west coast and Tapanuli: Annual report for 1927], 1928, unpublished, p. III, no. 4334; and A. F. A. M. Hendrichs, *Boschwezen Buitengewesten, Dienstkring Sumatra's Westkust, verslag over het jaar 1937* [Forest Service of the Outer Provinces, West Sumatra Forest District, 1937 annual report], 1938, unpublished, p. 23 (NI), no. 4303, both in Van Goor (1982).

29. My account of the social dynamics and responses of West Sumatra's first wave of plantation expansion draws on Kahn (1993, 187–223; 1984, 297–99) and Stoler (1986, 140).

30. Mintz (2013).

31. This is a trajectory of development with a basic blueprint that has been repeated worldwide. Compare Wolf and Mintz (1957), Dove (1983), Wells (1996), and Guthman (2017).

32. Peter Vandergeest and Nancy Peluso (2006) elaborated the dynamics of this late high-modernism wave of enclosures.

33. Lansing (1991, 12).

34. Peluso (2011).

35. Rachman (2011, 45–47).

36. Li (2007).

37. The victims of the New Order killings of 1965–66 were overwhelmingly leaders and members of the Indonesian Communist Party and its affiliated organizations, especially the Indonesian Peasant Front (Barisan Tani Indonesia) and the Plantation Laborers' Union of the Indonesian Republic (Sarekat Buruh Perkebunan Republik Indonesia). These were poor and lower middle-class farmers and laborers, living in villages, on plantations, and on the outskirts of urban centers (Robinson 2018, 122).

38. McGuire and Hering (1987, 222).

39. Ruzika (1978, 10), cited in Rachman (2011).

40. Afrizal and Patrick Anderson (2016) detail how the New Order enclosures were enacted in West Sumatra. Nancy Peluso (1992) provides an analysis of a similar process in Java.

41. Lucas and Warren (2003, 97).

42. Afrizal (2007, 48).

43. "Aren, the Beautiful Girl Being Ogled Again" (*Gadis Cantik Aren dilirik Lagi*), *Singgalang Independen*, 16 November 1989 (accessed at Singgalang Office Archives in May 2016).

44. "The Barren Lands of Aren Are Made Green Again" (*Tanah-tanah gundul di Aren dihijaukan kembali*), *Haluan*, April 1973 (accessed at Haluan newspaper office archive in May 2016).

45. In a series of articles published over three decades, Michael Dove elaborated how the competing views of grasslands in Indonesia between and among smallholders, agribusiness, and the state can only be understood by asking who benefits from either supporting the grasslands' growth or working to eliminate them. Smallholders often consider grasses as integral components of agroforesty, while agribusiness and state development agencies conceptualize them as wastelands (2019, 2004, 1986, 169–70).

46. A generation of environmental anthropology in Indonesia and Southeast Asia critiqued the "slash-and-burn" discourse as a way of maligning people living on the land (e.g., Dove 1985; Brosius 1999; Tsing 2005, 174). Slash-and-burn is just one name for a kind of agriculture known more neutrally as swidden agriculture, in which farmers shift cultivation over time between fields distributed across a forest area. The key variable in the sustainability of swidden agriculture is the length of the fallow period, wherein an individual patch is left to recover under only lightly managed regrowth conditions. If a fallow is sufficiently long, soil nutrients will be replenished, and the plot will sustain another rotation planted under annual crops (typically mountain rice or corn in Indonesia).

47. Beversluis, *Dienstkring Sumatra's Westkust*, no. 4025.

48. W. C. R. Schnepper, *Dienstkring Sumatra's Westkust: Houtvestersressort Fort de Kock. Maandrapporten Juli, Augustus en September*, no. 4573 [West Sumatra Service Section, Bukit Tinggi Forestry area: Monthly reports July, August and September], 1925, in Van Goor (1982).

49. Spencer (1966).

50. I draw instead from Harold Conklin (1957) to consider swidden agroforestry to be an ecologically grounded process that can support high levels of biological diversity and productivity beyond those of any industrial use. Almost a decade before Spencer's work, Conklin published *Hanunoo Agriculture: A Report on an Integral System of Shifting Cultivation in the Philippines*. Using early methods of environmental anthropology, Conklin detailed the complexity, productivity, labor efficiency, and ecological attunement of swidden-fallow agroforestry. Although slow to catch on, Conklin's research led the shift away from seeing forest cultivators as damaging agents of deforestation that predominated in colonial and post-colonial thought.

51. For example, see "Four People Detained" (*Empat Orang Diamankan*), *Suara Merdeka*, 1 May 1997.

52. "Alleged Mastermind of Timber Theft Detained" (*Diduga Menjadi "Otak" Pencurian Kayu di Cipada Oknum Pejabat Kehutanan Ditahan*), *Indo-News*, 26 August 1996.

53. "Flood . . . Because of Natural Karma" (*Banjir...Akibat Karma Alam*), internet chat-board communication, 6 December 2000 (accessed at Apakabar Archive, October 2015, www.library.ohio.edu/indopubs/).

54. "HPH [Industrial Logging Concession]: Forest King or Forest Pig" (*HPH: Raja Hutan atawa Babi Hutan*), internet chat-board communication, 7 April 2001 (accessed at Apakabar Archive, October 2015, www.library.ohio.edu/indopubs/).

55. Ramli, interview with author, February 2, 2016, as he was out checking on the cacao crop that he planted in the burned-out pine monoculture.

56. One case of the burning by "unknown origins" is reported in "Pine Forests Burn" (*Hutan Pinus Terbakar*), *Singgalang*, 22 October 2015.

3. THE PLANTATION LIFEWORLD

1. CIA Intelligence Advisory Committee, 5 March 1957, *The Situation in Indonesia*, Central Intelligence Agency Freedom of Information Act Electronic Reading Room Archive.

2. Republic of Indonesia Ministry of Information, 1958, Bandung and Djakarta People Condemn Padang Rebellion, Special Release no. 17, Leiden University KITLV Collection.

3. Republic of Indonesia Ministry of Information, 1958, Normalization of Freed Areas, Special Release no. 26, Leiden University KITLV Collection.

4. CIA Office of National Estimates, 2 May 1958, The Communist Position in Indonesia, memorandum for the director, Central Intelligence Agency Freedom of Information Act Electronic Reading Room Archive.

5. Peebles (1997) carried out long interviews with the leaders and engineers of the Corona project to provide a detailed institutional history of its creation.

6. Casiavera's smallholders were part of an earlier, Sumatra-wide reclaimers' movement of the 1950s. The Japanese retreat empowered hundreds of thousands of dispossessed smallholders and one-time plantation laborers to create homesteads in the colonial-era plantation estates and forest reserves. The mobilizations networked peasant unions, legal aid groups, and political parties. Increasingly motivated and organized, these movements challenged the old order of corporate land control. Independence dreams of a dominant smallholder political economy were in sight until pro-West, pro-plantation influences within the Sukarno administration nationalized the land concessions in the late 1950s, ending the smallholder reclamation movement and reinstating the industrial plantations. The army enforced this plantation regime with the arrest of some fifteen thousand reclaiming movement members by 1960 (Stoler 1986, 126–37; 1985, 142).

7. Abdul and Jon, *Collective Land Information and Conclusions* [*Keterangan dan Kesimpulan Tanah Ulayat*] (self-published, 2012). Abdul gave me a copy of the handout in 2016.

8. On multiple occasions in 2015 Abril and I walked the collective land while he narrated his memories of how the land was used throughout his life. I took a number of walks through the land with Yono, who recounted similar histories October 2015. A discussion forum with Casiavera's older residents on February 9, 2016, confirmed this historical ecology of the land.

9. Zain, interview with the author, October 25, 2015, while he drove us across Aren in his pickup truck. Zain is known to carry a binder with him at all times with print-outs of many local government regulations. He uses these documents to reinforce his arguments during formal meetings with local government officials and during more informal conversations with Casiavera's residents, which are common, spontaneous, and often heated enough that he feels the need to be able to refer to these regulations at any time.

10. Robert Shaplen, letter from Indonesia, *New Yorker*, 1 April 1974.

11. Yono, interview with the author, September 22, 2015, Yohanes's home.

12. Reclaimers made these remarks to me on September 2, 2015, during a nighttime conversation with a group of women in Casiavera after they wrapped up their weekly savings club meeting.

13. Komar, interview with the author, October 14, 2015, in his plot on the collective land. That Komar spent a decade in the army might have everything to do with why he had no interest in speaking about Casiavera's political history.

14. Agoez, interview with the author, October 16, 2015. We spoke while sitting on an old, handmade bench in his home garden, the wood planks of the bench long-ago worn smooth by use. Agoez was wearing a government-issued safari suit, a hand-me-down from his son, who works in the provincial planning and development ministry.

15. Where I join Timothy Mitchell (2002) in seeing silence as a sign of domination, James Scott (1985) sees it as an act of resistance meant to provide those that remain silent a strategic advantage in support of their freedom.

16. Arbi, interview with the author, July 30, 2015. I met him during a gathering at Casiavera's mosque. He lived on the far side of the Aren volcano, a more remote region known for its waterfalls and gibbons. He came to the gathering because his wife's brother was married to a woman from Casiavera.

17. This mother made this comment to me on April 25, 2016, as one of those awkward combinations of real concern and joking in her home just below the collective land.

18. Sudirman, a family-lineage leader of one of the families who lives along the border of the collective land, narrated the twentieth century of the land for me, as he lived it, at his home on October 29, 2015.

19. Group discussion in a Casiavera coffee shop on October 13, 2015.

20. Abril told me about his work as an overseer while we chatted in a coffee shop just below the collective land on April 16, 2016.

21. Zed, interview with author, April 12, 2016. Zed often visited my home in Casiavera to tell me about his transition from New Order policeman to active reclaiming movement member, a process Zed often spoke about in terms of a personal awakening to the meaning of justice for himself and the community he is a part of.

22. Agoez, interview with the author, October 16, 2015, at Agoez's wife's home. As I came to get to know Agoez and his family, I ended up spending more time with his son, Bayu, a peasant union member with a plot in the collective land who made most

of his income as a carpenter, a trade he learned during his four years working as a contract construction worker at Indah Kiat, Sumatra's largest pulp and paper mill.

23. Unfortunately, I was never able to speak with Monica Jessica Teuling. The few details of her life come from the public record required by her registration as a provincial legislative candidate: *Daftar Riwayat Hidup Bakal Calon Anggota DPR*, April 16, 2013.

24. Daud recounted his experience at the plantation at his wife's home in Casiavera on February 12, 2016.

25. Agoez, interviews with the author, October 13 and 16, 2015.

26. Wangga, interview with the author, February 13, 2015, while weeding between her bananas on her plot on the collective land.

27. See Mintz (2013).

28. Abril, interview with author, April 16, 2016.

4. FROM DISSENT TO OCCUPATION

1. With the idea of reclaiming, I seek to complement agrarian studies' typical analysis of land control. The explosion of research into agribusiness "land grabs" has provided important advances in understanding the ways that the powerful are able to make money from labor and land. Instead of ending my analysis with a critique of the land grab, however, I provide a prospective on smallholders' reclaiming with the hope of spurring new land reclaiming everywhere.

2. Aren, interview with the author, October 2015, at Aren's home.

3. Casiavera Community Council, Request That PT Dona's Application for Permission to Extend Its HGU on Casiavera's Customary Land Should Be Rejected (*Mohon Agar Permintaan Izin Perpanjangan HGU, PT Dona atas Tanah Ulayat Nagari di Casiavera supaya ditolak*), 24 December 1996, Wibawa's archive.

4. Provincial Head of the National Land Agency, The Problem with HGU No. 1/ The Elders of Casiavera Demand Its Cancellation (*Permasalahan Hak Guna Usaha No. 1/Desa Casiavera dituntut untuk dibatalkan oleh Ninik Mamak/Masyarakat di Kenagarian Aren, Prop. Sumatera Barat*), 1 October 1998, letter addressed to the Head of the National Land Agency, Jakarta, Wibawa's archive.

5. Protest mobilizations against Kedung Ombo, a World Bank dam project on Java, and Tanah Lot, a capitalist crony mess of a tourist resort and golf course in Bali, are the best-known early 1990s Indonesian social movements to combine environmental and land rights concerns.

6. I have never seen an original copy of this calendar. *Tempo*, 25 May 1991, 85; Lucas (1992, 79–80); and Bachriadi (2010, 124) all cite and provide reprinted images of it.

7. Dianto Bachriadi, leading scholar of agrarian justice, longtime activist, and the vice-chair of the Indonesian government's National Human Rights Commission, calculated these numbers using KPA's records (2010, 62–63).

8. Quoting Sudibyo (2001, 190), in Bachradi (2010, 384).

9. Lucas (1992).

10. Casiavera Community Council, Respectful request for a photocopy of the HGU and map of the collective land that is currently cultivated by Dona Company (*Mohon minta photo copy Warkah HGU, Peta tanah Ulayat Nagari Casiavera yang saat ini digarap PT Dona*), 29 February 1996, letter addressed to Governor of West Sumatra, Wibawa's archive.

11. Casiavera Community Council, Respectful request for a photocopy of the HGU and map of the collective land that is currently cultivated by Dona Company.

12. I do not have a copy of the Land Agency reply. I was only told that this was the first Land Agency response. The second letter from the council is Casiavera Community Council, Respectful Request for a Photocopy of the HGU on the Collective Land That Is Currently Cultivated by Dona Company (*Mohon minta Foto Copy Warkah HGU Tanah Ulayat Nagari Casiavera yang saat ini digarap PT Dona*), 9 December 1996, letter addressed to the Provincial Head of the National Land Agency, Wibawa's archive.

13. Casiavera Community Council, Update and Request to Reject (*Pemberitahuan dan Mohon untuk Pemblokiran*), 17 April 1997, letter addressed to the head of West Sumatra National Land Agency Office, Wibawa's archive.

14. In 1997 the council penned a collective statement called *History of the Collective Land* to justify Casaivera's reclaiming movement based on ideas of historical use, their ongoing "request and demand" for the return of the land, and ongoing need for control of the land to achieve community well-being (see Appendix I).

15. Casiavera Community Council, power of attorney (*Surat Kuasa*), 3 March 1997, letter, Wibawa's archive.

16. Leader of Bakorstana West Sumatra, Casiavera Customary Land That Was Taken by Dona Company Is Not State Land (*Tanah Ulayat Casiavera Digarap Oleh PT Dona Yg Bukan Tanah Negara*), 5 February 1997, Wibawa's archive.

17. Robinson (2018, 63).

18. O'Brien (1996).

19. County Head's Office, Meeting with Parties Involved with Dona Company (*Pertemuan Dengan Pihak PT Dona*), 12 December 1996, letter addressed to the Casiavera Community Council; and County Head's Office, Consultation Results Report (*Laporan Hasil Musyawarah*), 14 December 1996, letter addressed to the Aren District Head, Wibawa's archive.

20. Tia, interview with the author, October 12, 2015. Tia is from one of the wealthiest families who live on the upper flanks of the volcano. Her mother owns three hectares of irrigated paddy, which her family both cultivates and rents to sharecroppers. Still, her family was committed to the reclaiming. Tia told me that more than ten of her closest kin joined her that day at the blockade.

21. The role of elite women in Minangkabau is not limited to land relations, of course. Most scholars of the Minangkabau territories see women as having fundamental influence in most economic activity in the region, even while the dominant

political economy is structured along patriarchal capitalist and state structures (e.g., Blackwood 2000; Sanday 2002).

22. Zed, interview with the author, April 12, 2016, at the author's home in Casiavera. Zed retired from his post as a policeman to become mayor and provide leadership to the movement. 'I had to [retire], how could I both be a policeman and support the people here, when it was how it was?'

23. Reports and analysis of violence in Sumatra's agrarian spaces during the New Order were rare, for censorship and repression of such research was ubiquitous. The end of the New Order brought a reduction of these risks and a great growth of research on Indonesia's violent environments (e.g., Hadiz 2000; Anderson 2001; FERN 2001; Peluso and Harwell 2001; Human Rights Watch 2003).

24. Land Rights Staff, West Sumatra National Land Agency, and Casiavera Community Council, Conclusion from Meeting with the West Sumatra National Land Agency (*Kesimpulan Pertemuan Dengan Kantor BPN Propinsi Sumatera Barat*), 6 January 1997, Wibawa's archive.

25. Casiavera Community Council, Update and Request for Rejection.

26. Casiavera Community Council, Update and Request for Rejection.

27. Casiavera Community Council, Update and Request for Rejection

28. Casiavera Community Council, Update and Request for Rejection.

29. Provincial Head of the National Land Agency, Problem with HGU No. 1.

30. Provincial Head of the National Land Agency, Problem with HGU No. 1.

31. Zed, interview with the author, April 16, 2016, at the author's home in Casiavera.

32. The visionary Italian autonomists have located movement power with the multitudes since the 1970s (Virno 2004).

33. Eric Hobsbawm (1974) provides a definitional account of the land occupation as protest.

34. I calculated this rough estimate of reclaiming by using Indonesian Peasant Union publications, Sumatran newspapers, and reports from the Consortium for Agrarian Reform and Oil Palm Watch. Reclaimers from more than sixty different villages have occupied some twenty-five industrial plantation concessions.

35. Sirait (2015, xvi, 28, 98, 106, 138, 176).

36. See Li (2014).

37. My account of the Pawartaku reclaiming draws primarily from Wijaya (2018).

38. Human Rights Watch (2016) and Rahman (2018) detail the repression of the Fajar Nusantara communities.

39. The centrality of the smallholder in post-colonial Indonesia was made certain in President Sukarno's famous 1959 speech, "Political Manifesto." In it, the president declared the government's support of new land reform that would return the European-operated plantation lands to the tiller. Sukarno justified the reform with a series of claims about the importance of peasants, including that smallholders who owned land were the most effective agricultural producers, landlords caused wasteful

use of resources, and increasing tillers' ownership of the land would increase equality and prosperity across the nation.

40. Jeffrey Hadler (2008) most clearly framed Minangkabau's outsized role in the formation of the Indonesian Republic as the wonderful result of the tensions and frictions of competing influences of the matriarchate, Islam, and post-colonial geopolitics.

41. Wiradi (2000, 117).

42. Lubis (2013).

43. Wijardjo and Perdana (2001).

5. ORGANIZING THE MOVEMENT

1. Afiff et al. (2005), Peluso, Afiff, and Rachman (2008), and Bachriadi (2009, 2, 22) provide detailed accounts of the post-Suharto smallholders movement in Java.

2. For more on these, the most significant agrarian movements in Indonesia since the early days of the founding of the republic, see Aditjondro (1998), Rachman (2002), Bachriadi (2010, 386), and Warren (2013, 243–45). Other English-language scholarship has focused less on occupations and reclaiming as such and more on the Indigenous aspects of these movements (cf. Tsing 1999; Li 2001; Astuti and McGregor 2017).

3. See appendix II for my translation of the 1998 Indonesian Peasant Union Charter document from the original Indonesian.

4. SPI is only one of many national peasant unions that all trace their origins to this time. Others include National Peasant Alliance (Aliansi Petani Indonesia), National Peasant Union (Serikat Tani Nasional), and Agrarian Reform Movement Alliance (Aliansi Gerakan Reforma Agraria). These operate alongside advocacy organizations like Consortium for Agrarian Reform (Konsorsium Pembaruan Agraria), political parties, and policy institutes. All of these organizations emerged as part of a new generation of social movement organizations around the world that took up land occupations and reclaiming (Rosset 2006, 221–23).

5. Reclaiming and occupation movements were joined with a surge in less-organized, more individualistic kinds of agrarian energy. The ongoing "monetary crisis" (*krisis moneter*) spurred migrants to leave Java's urban areas and settle Sumatra's Bukit Barisan to produce coffee, one of the few forest commodities that held its regional value during the recession (Sunderlin and Resosudarmo 1999). Overall, the surge in the forests was largely spontaneous and unmanaged by any movement organization. Reclaimers established new land claims alongside longtime squatters. Timber mafias flooded the forests, getting rich quick.

6. Li (2014, 84).

7. Yohanes, interview with the author, October 13, 2015. I often met Yohanes as he walked from his home in the center of the collective land down into the village, to gather at the coffee shop or mosque. He was one of the few to walk instead of driving a motorbike.

8. Sanday (2002, 15).

9. For this account I draw on a series of interviews I carried out in 2015–16 with the four men who were elected mayor of Casiavera from 1995 to 2016, three members of the council, and several residents who claimed plots in the collective land during the first wave of occupation.

10. I have a copy of the document Casiavera Council Regulation Number 1, Regarding the Management of the Collective Land (*Peraturan Nagari Casaivera Nomor 01 Tentang Pemanfaatan Tanah Ulayat Nagari*), signed by the head of the Casiavera Council in 2003. Copies of the regulation are held in the council offices.

11. For a review of similarly strict limited access regimes, see Netting (1993, 30–35). The social relations of the reclaimed land can be compared with the British commons system where access was preferentially enjoyed by the peasants of highest status. Many households that lived within common grazing lands did not have standing to use them, usually because they were not considered 'original' members of the community (De Moor, Warde, and Shaw-Taylor 2002).

12. Sumatra's criminal underworld includes many who look and dress like Susilo who take the label *preman*, men who identify as half opportunistic criminal and half folk hero. Many *preman* turn to ruthless extortion of those less willing or able to use force to get what they want, often at the behest or employ of the plantation land elite who use *preman* as underground enforcers of their land grabs. *Preman* are also employed to disrupt and attack leftist workers', agrarian, and environmental movements (Gilbert and Afrizal 2019).

13. The tendency for some anthropologists, geographers, environmentalists, and activists to think that all local forms of governance will result in democratic societies has very much faded, but a romanticization of the local still persists. Wealth inequality, sexual exploitation, xenophobia, and outright slavery are just three common dangers of localism (Gadgil and Guha 1993; Mohan and Stokke 2000; Harvey 1993).

14. Daud, interview with the author, February 12, 2016, at Daud's home on the collective land. Daud's role as manager of the Casiavera mosque and regional SPI organizer is significant: he spans a divide that was once devastatingly deep. During the Cold War anti-leftist genocide, Muslim parties supported the bloodshed. They considered their position antagonistic to the left. Daud provided an indication that these constituencies can be bridged. Daud told me he is careful to not press the issue. He does his best not to mix SPI and mosque activities, he told me. He does not hold organizing meetings at the mosque or talk politics during Quran readings, for example. But by staying active in both causes he demonstrates to others that there is nothing difficult about reconciling a demand for peasant rights with his spirituality and organized religion. During the land occupation, Daud did draw on his connection to the religious community, however, bringing in the mosques' mainstream local political influence based on a moral appeal to the rights of Casiavera's residents.

15. Fortmann (1985), Fortmann, Antinori, and Nabane (2010), Rocheleau and Edmunds (1997), and Schroeder and Suryanata (1996) all see the social power of tree planting, with multiplicitous effects on social relations of gender, equality and hierarchy. Tania Li's (2014) work in Sulawesi shows the way that tree crop booms can cause

Indigenous smallholders to be dispossessed from below by their own families and neighbors, as successful cacao planters and traders acquire and privatize lands of less successful tree planters, who can be pushed over the brink of exhaustion and starvation.

16. Yunus's complete agroforestry plot was a place he could expound to me his ideas of the smallholder economy and relation to the land. We discussed ideas of agroforestry on the collective land in detail on October 6 and 9, 2016. This particular idea, that tree crops should not be subject to regulation from the state or Minangkabau customary authority, went against centuries of just such regulations in West Sumatra. Yunus's conflicting view was not born of naivete, because Yunus had long played an active role in the discussion and history of customary authority on the volcano, but out of smallholder dissent against the tendency for higher authority to intervene in his agroforest plot.

17. Arif's comments are recorded in an Indonesian Peasant Union member newsletter, documenting the weeklong member visit to Casiavera in January 2012. Arif gave me a copy of the newsletter during the second of my three interviews with him in his home in 2016, outside of Padang.

18. Sanday (2002, 21).

19. I learned of this event first through an article in the local paper: "Kantor KAN Casiavera Disegel," *Padang Ekspres*, August 8 2015. I spoke with the chairman of the customary council and one of the dissenting elders (*ninik mamak*) a few days later for a fuller understanding of the situation, when I saw them on the collective land.

20. Sanday (2002, 29–33).

21. Of the anarchist scholars, James Scott (2009) has achieved the most mainstream consideration. Scott draws on a long lineage of anarchist thinkers who all, to varying degrees, made contributions to debates on dynamics of power between society and the state. See, for example, Bakunin ([1873] 1971), Kropotkin (1902), Goldman (1911), and Ward (1973).

22. Malaka ([1948] 1991, 8).

23. Reid (2005, 64).

24. I draw on two long conversations I had with Nurul in the evenings of August 8 and 11, 2015. We chatted in the joint office and home of a local politician and environmentalist in Bukit Tinggi.

25. For example, Dt. Nago Besar, Minangkabau Indigenous scholar, summarized the origins of the matriarchate for Peggy Reeves Sanday thus: "'Where can the child find food and land, titles, and home (*pusaka*)? Like growth in nature, we always know from whom the child descends: the mother'" (Sanday 2002, 25).

26. Nurul's comment on August 11, 2015, was the first time I had heard people refer to Casiavera's specific social relations as "dense." After spending many months more in Casiavera, I came to hear discussion of this "density" several times in relation to Casiavera's reclaiming.

27. One social dimension where the equality of Casiavera's polity begins to break down is age. The council and other family lineage leaders are all over forty. Younger

adults and teenagers have no formalized role in deliberations. While class and gender divisions in power are more attenuated, this polity tends toward gerontocracy.

28. The council's shift challenges James Scott's (1976) idea that these forms of resistance are the result of peasants falling through the subsistence floor into desperate poverty. In this line of thinking, having reached the point at which the subsistence survival of the peasantry was threatened, they would rebel. Faced with starvation, resistance became a risk worth taking. But in Casiavera the reclaimers' position was not so dire. Casiavera's reclaimers were not starving, although many were impoverished. They mobilized facing great risk without being pushed to total destitution. Kartodirdjo (1966) gives a fascinating account of how during the colonial era Javanese peasants similarly transitioned from fearful obedience to open revolt without waiting for the utter failure of their subsistence strategies.

29. My observations and interviews on collective land use did not include all land users. I did walk all the plots, determining that nearly all plots were being used, but I was not able to determine every plot user.

30. Von Benda-Beckman and von Benda-Beckman have considered the reconstitution of Indigenous Indonesian governance systems in detail (e.g., 2013, 39–60 and 2011, 177–80).

31. Eric Wolf (1957) was one of the first academics to publish this idea.

32. Daud spoke with me about the issue as we ate lunch at his home on February 2, 2016.

33. I joined this meeting on April 17, 2016.

34. The trader told me his ideas about the land conflict on April 20, 2016, while we drove to the market to sell a vegetable crop there.

35. I have a copy of the Aren District court's decision, as well as arguments and testimonies submitted during the trial and field inspection. These documents are stored at the Aren District Court as Civil Case No. 6/PDT.G/2012/PN PYK. There are three collections of documents, dated 2 October 2012, 4 October 2012, and 16 October 2012.

36. Civil Case No. 6/PDT.G/2012/PN PYK.

37. I have a copy of the document submitted to the Aren District Court. Conclusions of Defendants I, II, and III, 2 October 2014, prepared by the attorneys Nusantara, Oktavia, and Siregar.

38. This expiration date is given in *Daftar Isian Identifikasi Dan Penelitian Tanah Terlantar Hak Atas Tanah* (List of Identification and Research on Land Rights on Neglected Land), 22 January 2010, National Land Agency District Office Report, Padang, West Sumatra, Wibawa's archive.

6. DIVERSIFYING THE LAND, 1998–2016

1. Over several days in the fall of 2015 and again in spring 2016, I worked alongside Wangga in her plot. During our work Wangga commented often on how she had transformed the degraded land into her agroforest. Discussion of these changes often

led to her thoughts about her own contemporaneous transformation, from landless food service worker to agroecologist.

2. I recorded these details of Wangga's life while working with her in her plot in the collective land on October 16, 2015 and while sitting in the coffee shop alongside her home, which she runs in the evenings, on May 3, 2016.

3. This conversation took place on the collective land on February 2, 2016. Rice was an especially gendered crop in Casiavera. Mostly it was women who owned paddies. And mostly it was women who worked to harvest rice in reciprocal work groups. When a woman serves rice from her own paddy in the home, it is not a family resource held in common. No, it is *her* rice.

4. As the expansion of industrial agriculture and its associated carbon emissions has increased in the Bukit Barisan since 2000 or so, agroforests have taken on a greater importance in carbon storage as the only current Sumatran agricultural system capable of significant uptakes of atmospheric carbon (Villamor, Pontius, and Noordwijk 2013). Agroforests act as a carbon sink (Steffan-Dewenter et al. 2007). Agroforest cultivation stores about twice as much carbon as oil palm or rubber monocultures per unit area (Sandker and Suwarno 2007). In Indonesia, agroforests store more carbon than uninhabited forests (Carlson et al. 2012). As land pressures have increased, more and more industrial plantation expansion targets agroforests, making Indonesia one of the top ten national contributors worldwide to greenhouse gasses (Indonesian Ministry of Environment 2010).

5. Satellite photographs and the information about land use they contain are fundamentally a form of mapmaking, and the powerful have used mapmaking as a tool to control territory for centuries. Any use of spatial technologies, of these "master's tools," requires an explicitly self-reflexive approach. Central to this approach is the need to create maps of change that refuse the tendency of mapmakers to erase Indigenous and smallholder presence and fix the landscape in time (see appendix III).

6. While uptake of Connell's revisionist ecological concept was slow, a post-equilibrium theory of diversity has finally become an accepted, if not majority, position within ecological science (Molino and Sabatier 2001).

7. Connell (1978). In West Africa, the work of James Fairhead and Melissa Leach (1996) showed that smallholders cultivated forests to support their reproduction. Smallholders did not, as colonists and the post-colonist government and environmentalists claimed, destroy them. With a similar conclusion in Central America, Susanna Hecht and Sassan Saatchi (2007) described a flourishing regrowth of the forests at the hands of smallholders after they were given access to state lands.

8. Hariyadi and Ticktin (2012), Aumeeruddy and Sansonnens (1994), and Michon, Mary, and Bompard (1986) provide careful accounts of West Sumatran agroforestry.

9. A detailed study of the relative abundance and species richness of birds in Sumatra in agroforestry and uncultivated "primary" forests shows that there are important distinctions between coffee, durian, and damar resin agroforests, and none of these agroforests are equivalent to uncultivated forests. Most noticeably, the largest

frugivore and insectivore bird species were not able to survive in the agroforests and depend completely on uncultivated fruit tree species found in the wild forests for a place to live (Thiollay 1995).

10. After generations of exclusion from academic research efforts, over the past three decades agroecology has finally gained a position in Western scientific thought. Universities in Latin America and Southeast Asia were earlier than those in North America and Europe to bring in agroecology. One of environmental anthropology's greatest contributions is the clear demonstration that for many smallholders in many landscapes, agroecology is preferable to industrial agriculture across many dimensions, including human health, economic risk, labor efficiency, agricultural yields, and sustainability (Altieri and Toledo 2011; Altieri 1995, 112, 206; Netting 1993).

11. The two pieces of machinery that were common in Casiavera were the gas-powered backpack weed cutter and a mechanical coconut scraper. The first was used to protect young tree and vegetable crops from being overgrown. It arrived around the year 2000. The second women used to make coconut flakes and milk, replacing long hours at the coconut scraper stool. Perhaps the weed cutter became popular for the same reason as the coconut stool: it made weeding and scraping—two of the more tedious smallholder tasks—less tiring and quicker.

12. "While corn ricks and other property was destroyed (as well as some industrial machinery in country districts)," writes E. P. Thompson of a laborers' revolt in 1830, "the main assault was on the threshing-machine, which patently was displacing the already starving laborers. Hence the destruction of the machines did in fact effect some immediate relief" ([1963] 2001, 27). Eric Hobsbawm (1952) wrote about the luddites directing their ire at machines because of their anger at the capitalist system that makes these machines material.

13. Altieri (1995, 33, 248) and Altieri and Toledo (2011).

14. Pollen and biomass burning records indicate productive forest management in West Java approximately sixty-five thousand years ago, shortly after the arrival of humans (van der Kaars et al. 2001). By the Early Holocene about twelve thousand years ago, there is a clear record of humans transporting and propagating tree crops like sago and sugar palm across wide areas of the Sunda Shelf encompassing Sumatra, Java, and Borneo (Hunt and Rabett 2014). Pre-colonial agroforestry remade the forests and food economies of the Sunda in a way that blurred the distinction between wild and cultivated forests (Hunt, Gilbertson, and Rushworth 2012).

15. Scott (1999, 335–36) argues that the arts of domination require Indigenous metis to be erased to impose discipline and gain taxes. Vandana Shiva (1993) popularized the phrase "monoculture of the mind" to convey the social simplification of industrial development.

16. This interaction took place on October 18, 2015. A few weeks after our outing, Midwal was caught in bed with his married neighbor; he took part in a series of community restorative justice meetings attended by hundreds of people. At the end of the meetings, Midwal and his lover apologized to her husband, and they agreed to

pay the husband fifty sacks of dry concrete and marry, which they did the same day that the neighbor got divorced from the husband. After all of this, many men and women asked me not to spend time with Midwal in the forests, as a form of punishment for this behavior that they did not agree with, a suggestion I mostly abided by.

17. Polong's (2010) unpublished essay is *Pembangunan Diklat Permakulture Basis SPI Talang Keramat (Sumsel): Solusi untuk Perubahan Iklim* (May 4, 2010); provided to me by the author.

18. On the entanglements of ecological diversity and human affect, place, enskillment, and resilience, see Perfecto, Vandermeer, and Wright (2009) and Ingold (2000, 47, 55).

19. See the Javanese volcanic cosmologies (Dove 2008).

20. I spoke with this forest farmer on September 24, 2015, as he was cutting elephant grass in his agroforest plot on the collective land as fodder for his penned cattle.

21. I met with Surya several times in July 2015. The comments I detail here came from a discussion on July 29, 2015.

22. Altieri and Toledo (2011) argue that such is the case in Latin America. The reclaiming in Casiavera adds to evidence that agroecology's revival has reached Indonesia as well.

23. The bricolage understanding of rural work has done much to dissolve the rural/urban ontology. Marc Edelman (1999, 204–8) lays out some of the political dimensions of these livelihoods as they relate to peasant autonomy. The economic and ecological dimensions of bricolage livelihoods are just as significant; see, for example, Kizos et al. (2011), Brookfield and Parson (2007), Rigg (2000, 204–9), Marsden (1990), and Gasson (1986).

24. Bricolage has grown in the US countryside as well. Brookfield (2008) shows that in the United States, by the 1980s agriculture provided only one-half of farm family income. In the European Union, by 1987 one-third of farmers had "other gainful activities." Across Latin America at the end of the 1990s, more than two-fifths of the rural population had their principal occupation away from the farm (Kay 2008).

7. THE PREDATORY WORK THAT REMAINS

1. It is difficult to determine the extent of these land transfers, as people using the land knew such transactions were a reason for the council to impose sanctions or to revoke use of plots on the collective land. A few reclaimers told me that certain influential men were profiting from plots actually registered under extended family members, but I was not able to confirm or rule out these transfers. Over the months I spent on the land, it became clear to me who worked which plot and if they were the registered user of the plot, but it was more difficult to know if they were working for themselves or for the profit of others.

2. Reid (1985) believes that by the end of the late 1700s Casiavera and the surrounding region was the center of Sumatra's valuable tobacco industry, all produced

by smallholders (see also Marsden 1811, 323). Deli, now known as Medan, became the center of industrial plantation tobacco by the 1850s (Boomgaard 1999, 59).

3. Such "bosses" have played a prominent role in Sumatran tobacco since the nineteenth century, when wealthy merchants began a system of extending credit to both producers and middlemen to ensure their access to the crop to sell to the international fleets that regularly arrived at Sumatra's ports in search of the plant (Boomgaard 1999, 59).

4. See Watts (1994), who tends to see debt cropping as a way for capitalists to transfer the risks of crop failure to the tiller. Some working in Southeast Asia, where sharecropping relations are often between kin and neighbors, see sharecropping as a typically benevolent relation that benefits the smallholder (Hefner 1990).

5. Lufti, interviews with the author, September 24, 2015, at Lufti's cattle shed on the collective land that he manages with an agricultural cooperative, and April 18, 2016, at his wife's home, a fifteen-minute motorbike ride along the upper flanks of the Aren volcano.

6. Mansur, interview with the author, April 20, 2016, outside Mansur's trading house.

7. Contract farming became a mainstay of smallholder agriculture in sub-Saharan Africa by the 1990s (Little and Watts 1994). Contract farming is also widespread across Thailand, Malaysia, and Indonesia (Glover and Lim 1992, 4). In both regions monopsonist smallholder relations led to power imbalances and uneven distribution of profits. Reflecting on the situation in Java before the fall of the New Order, Ben White (1997) argues that contract farming in itself does not necessarily "spell hardship or doom for smallholders" but often did in practice because of how power was exercised in rural society.

8. I only interviewed the tobacco boss once, on April 27, 2016, when he was overseeing the harvest of a tobacco plot near my home in Casiavera.

9. Zain, interview with the author, October 25, 2015, while paying a reclaimer her share of the income she earned tending one of the cows Zain had recently sold.

10. World Bank (2018).

11. The persistence of sharecropping has long been a topic of debate in agrarian studies and rural economics. Writing against both the Leninist and neoclassical economic schools that saw sharecropping as impracticable under twentieth-century commercial agriculture, Cheung (1969) demonstrated the potential suitability of sharecropping for croppers, especially when their share of the harvest is proportional to their share of inputs invested in the crop. In Casiavera these conditions were not often met, with inequalities between cropper and landholder determining an exploitative relation.

12. Such agrarian differentiation has an intense and nearly omnipresent history in Sumatra and Java. Far from any sharing of poverty and "flaccid indeterminateness," as Clifford Geertz (1963) so incorrectly characterized Javanese society, sharp divisions among and between Indonesia's agrarian classes have played out as large-scale agrarian conflict involving agribusiness, smallholders, and the landless for decades (White 1983, 1997, 2016).

8. RECLAIMING SOLIDARITIES

1. Sukardi Bendang, Maintaining Customary Land for the Children and Grandchildren: Case History of the SPI group in Casiavera (*Mempertahankan Tanah Ulayat untuk Anak Cucu Kamanakan: Sejarah Kasus SPI Basis Casiavera*), 18 December 2012, Indonesian Peasant Union Report, Jakarta, SPI-West Sumatra Archive, Padang.

2. One well-known proponent of organic methods on the volcano, Ilyah, told me he had learned about the health and environmental importance of organic agriculture as a university student in Padang in the early 1990s. I spent a number of days staying at his homestead and farm at the base of the volcano in August 2015, where Ilyah shared with me his vision of how organic production methods were compatible with smallholder technology and knowledge, what he liked to call "smallholder equipment" (*peralatan petani*).

3. Two members of the Research and Advocacy Institute, interview with the author, August 20, 2015, in their offices, Padang, West Sumatra.

4. Malin, interview with the author, February 4, 2016.

5. Malin, interview with the author, February 4, 2016, in Malin's home in Casiavera.

6. My observations about SPI membership stem from visits to eight movement reclaiming sites in 2012 and 2015 in Sumatra, beyond Casiavera. Casiavera differs from other SPI reclamations in that nearly all Casiavera's laborers-turned-smallholders identify as indigenous Minangkabau, not Sumatran-Javanese. More common across Sumatra's other land occupations are groups of adult descendants of Javanese migrants who had come to labor in Sumatra's plantations, but who wanted something more for themselves than plantation life and set out to claim land for themselves.

7. Lufti, interview with the author, April 18, 2016, on the collective land, Casiavera.

8. Peet and Watts (1996).

9. Daud, interview with the author, February 3, 2016.

10. Saragih joins two dozen additional founding members of SPI who have visited with their MST counterparts over the years to study MST's diverse set of tactics of logistical support, ideological and agricultural training, legal analysis, advocacy, and media outreach.

11. I spoke with Luzon on many occasions in 2015 and 2016 at the cooperative's cattle shed. These comments were from a conversation on April 11, 2016.

12. I spoke with Mansur in his cooperatives' plot on April 21, 2016.

13. Malin made these comments to me on my second visit to Casiavera in August 2015 while we toured his dairy cooperative.

14. My account of Sino and Citra's life history is written from four interviews and a series of encounters and conversations with them, September 2015 to May 2016.

15. Susilo, interview with the author, October 1, 2015, on the collective land.

16. Eric Wolf's (1955) work, *Types of Latin American Peasantry*, is one of the original descriptions of the subsistence peasantry. A much more recent description of the economic relations of the modern peasant—landed, poor, and producing for subsistence but also profit—is given in Bernstein (2010). Where Wolf drew on the

Chayanovian concept of peasant economy that centered around subsistence production to reproduce the bounded peasant household, Bernstein acknowledged that few classical peasants still exist in the countryside. Virtually all modern "peasants" are smallholders who value self-provisioning but rely on the sale of agricultural commodities as well as their own labor for a wage for survival and social reproduction.

17. Tia made these remarks to me on the morning of September 18, 2015, while we worked the cooperative's plot.

18. There were at least five more active women smallholder cooperatives in Casiavera. Almost all had associated informal credit groups. I especially enjoyed joining these credit group meetings for the lively conversation and often moving conversations surrounding members' votes to allocate loans.

19. Daud, interview with the author, February 13, 2016.

20. Blackwood (2000, 180–83).

21. These observations and interview are from February 2, 2016, at Mansur's home and trading house in Casiavera. As his business has grown, he has added rooms to what was first what he calls his father's home, on his mother's inherited land in the village. In addition to the three-bedroom home and trading house, Mansur has built a small market stand along the side of the home that faces the dirt motorcycle track that leads to a small hamlet off the paved road. He also built his barber shop stand that is alongside the trading house, with bright compact florescent lights and two walls of mirrors, which he works in most nights.

22. Waranto told me his thoughts about Casiavera's economy in his home on April 16, 2016.

23. Dias Martins (2006, 269).

24. This first meeting took place in late 2015. I met with Aren several times after that through 2016.

25. Escobar (2018, 172–74).

26. Ortner (2014).

27. Compare Lenin ([1899] 1988) and Kautsky (1899).

28. Marx (1882), quoted in Stedman-Jones (2016, 587).

29. For example, Ward (1973) and Clastres (1977).

30. Armand ([1911] 2010).

31. Habermas (1987, 400).

CONCLUSION: GOING BEYOND

1. On the world-building potential of new economic imaginations, see Appel (2014).

2. Bear et al. (2015).

3. Li and Semedi's (2021, 12) ethnography of two oil palm plantations in Kalimantan reveals the corporate occupying force that accomplished the ongoing enforcement of plantation life. In the plantations where Li and Semedi worked, laborers and villagers were conscripted into plantation labor without mobilizing to remove the

corporation or local elites (143). Much, much more common across Sumatra and Kalimantan are plantations of conflict and contention, where dissenters—usually in the minority—work to undo or reform plantation life as they and their families live it.

4. See Archambault (2016), Gibson-Graham (2006), and Gupta and Ferguson (1997).

5. Rural studies are increasingly concerned with ruined industrial landscapes and the arts of survival emerging from them; see Wiley (2008), Tsing (2015), Ferguson (1999), Tsing et al. (2017), and Vaccaro, Harper, and Murray (2016).

6. See Majka and Majka (2000), Harrison (2008), and Holmes (2013).

7. My account of the eviction of the union members from the land is drawn from interviews with union members in 2012 and Jambi newspaper accounts of the eviction day. I have a cell phone video of the eviction and have spoken with two union members present that day.

8. I visited the bankrupt plantation in 2011, carrying out two weeks of interviews and observations with the Javanese migrants in the plantation, Malay squatters, and local indigenous families who had come to return to the land.

9. Edelman (1999, 206).

10. Rosset et al. (2011).

11. Toledo, Boege, and Barrera-Bassols (2010) and Hecht (2014).

12. Safransky (2017).

13. Simpson (2014).

14. Rosset, Patel, and Courville (2006).

15. Estes (2019, 248), echoing Simpson (2014).

16. Cf. Vaccaro, Harper, and Murray (2016) and Tsing (2015, 41).

17. Land grabs and more subtle forms of dispossession are nothing if not persistent through changing temporalities and political economies (Jegathesan 2015, 2019).

APPENDIX III. COUNTER-MAPPING

1. The National Land Agency map of the collective land appears in *Daftar Isian Identifikasi Dan Penelitian Tanah Terlantar Hak Atas Tanah* [List of identification and research on land rights on neglected land], 22 January 2010, National Land Agency District Office Report, Padang, West Sumatra.

2. I accessed the National Land Agency Online Map in September 2017 at http://peta.bpn.go.id/.

3. Rocheleau (1995, 2005) and Orlove (1991) both find maps are used primarily as a state biopolitical technology of control, negation, and erasure of rural people. Much critical work on mapping since then has underscored this interpretation of the use of maps. Nancy Peluso (1995) was one of the first Western scholars to give consideration to the emancipatory potential of mapping technologies in Southeast Asia.

4. Sikor and Lund (2009) and Harley (1989) consider the damage that accepting state administrative boundary maps as uncontested, exclusive knowledge does to struggles for land, resources, and sovereignty.

REFERENCES

Abdullah, Taufik. 1966. "Adat and Islam: An Examination of Conflict in Minangkabau." *Indonesia* 2: 1–24.

Aditjondro, George J. 1998. "Large Dam Victims and Their Defenders: The Emergence of an Anti-Dam Movement in Indonesia." In *The Politics of Environment in Southeast Asia: Resources and Resistance*, edited by P. Hirsch and C. Warren. New York: Routledge.

Afiff, Suraya, Noer Fauzi Rachman, Gillian Hart, Lungisile Nitsebeze, and Nancy Lee Peluso. 2005. *Redefining Agrarian Power: Resurgent Agrarian Movements in West Java, Indonesia*. Berkeley: University of California Berkeley Center for Southeast Asian Studies Working Paper.

Afrizal. 2007. *The Nagari Community, Business and the State: The Origin and the Process of Contemporary Agrarian Protests in West Sumatera, Indonesia*. Bogor, Indonesia: Forest People Program and Sawit Watch.

Afrizal and P. Anderson. 2016. "Industrial Plantations and Community Rights: Conflicts and Solutions." In *Land and Development in Indonesia: Searching for the People's Sovereignty*, edited by J. F. McCarthy and K. Robinson. Singapore: ISEAS—Yusof Ishak Institute.

Altieri, Miguel A. 1995. *Agroecology: The Science of Sustainable Agriculture*. 2nd ed. Cambridge: Cambridge University Press.

Altieri, Miguel A., and P. Rosset. 1996. "Agroecology and the Conversion of Large-Scale Conventional Systems to Sustainable Management." *International Journal of Environmental Studies* 50 (3–4): 165–85. https://doi.org/10.1080/0020723960 8711055.

Altieri, Miguel A., and V. M. Toledo. 2011. "The Agroecological Revolution in Latin America: Rescuing Nature, Ensuring Food Sovereignty and Empowering Peasants." *Journal of Peasant Studies* (July): 37–41.

Anderson, B. R. O. 2001. *Violence and the State in Suharto's Indonesia*. Ithaca, NY: Cornell University Press.

Appel, Hannah. 2014. "Occupy Wall Street and the Economic Imagination." *Cultural Anthropology* 29 (4): 602–25.

Archambault, Julie. 2016. "Taking Love Seriously in Human-Plant Relations in Mozambique: Toward an Anthropology of Affective Encounters." *Cultural Anthropology* 31 (2): 244–71.

Armand, Emile. (1911) 2010. "Mini-Manual of Individualist Anarchism." In *l'Encyclopedie Anarchiste*. Translation by Shawn Wilbur. www.theanarchist library.org/library/emile-armand-mini-manual-of-individualist-anarchism.

Astuti, Rini, and A. McGregor. 2017. "Indigenous Land Claims or Green Grabs? Inclusions and Exclusions within Forest Carbon Politics in Indonesia." *Journal of Peasant Studies* 44 (2): 445–66.

Aumeeruddy, Y., and B. Sansonnens. 1994. "Shifting from Simple to Complex Agroforestry Systems: An Example for Buffer Zone Management from Kerinci (Sumatra, Indonesia)." *Agroforestry Systems* 28: 113–41.

Bachriadi, Dianto. 2009 "Land, Rural Social Movements, and Democratisation in Indonesia." Transnational Institute, Rural New Politics—Rural Democratisation Research Project. www.tni.org/en/article/land-rural-social-movements-and-democratisation-in-indonesia.

———. 2010. "Between Discourse and Action: Agrarian Reform and Rural Social Movements in Indonesia Post-1965." PhD diss., Flinders University, Adelaide, Australia.

Bakunin, Mikhail. (1873) 1971. *Statism and Anarchy*. In *Bakunin on Anarchy*, edited and translated by S. Dolgoff. New York: Vintage Books.

Bear, Laura, K. Ho, A. L. Tsing, and S. Yanagisako. 2015. "Gens: A Feminist Manifesto for the Study of Capitalism." *Fieldsights*, March 30. https://culanth.org /fieldsights/gens-a-feminist-manifesto-for-the-study-of-capitalism.

Bernstein, Henry. 2010. *Class Dynamics of Agrarian Change*. Halifax, NS: Fernwood.

Bhandar, Brenna. 2018. *Colonial Lives of Property: Law, Land, and Racial Regimes of Ownership*. Durham, NC: Duke University Press.

Bjork-James, Carwil. 2020. *The Sovereign Street: Making Revolution in Urban Bolivia*. Tuscon: University of Arizona Press

Blackwood, Evelyn. 2000. *Webs of Power: Women, Kin, and Community in a Sumatran Village*. New York: Rowman & Littlefield.

Boomgaard, P. 1999. "Maize and Tobacco in Upland Indonesia, 1600–1940." In *Transforming the Indonesian Uplands: Marginality, Power, and Production*, edited by T. Li. Amsterdam: Harwood Academic Publishers.

Borras, Saturnino M. 2016. "Land Politics, Agrarian Movements, and Scholar Activism." Inaugural Lecture, International Institute of Social Studies, April 14. www.tni.org/en/publication/land-politics-agrarian-movements-and-scholar -activism.

Bos, J. 1911. *The Sumatra East Coast Rubber Handbook, 1911: A Manual of Rubber Planting Companies and Private Estates, Details As to the Present Stage of Development.* Manchester, UK: Messrs. Lintner and Co. Limited.

Bourgois, Philippe. 1988. "Conjugated Oppression: Class and Ethnicity among Guaymi and Kuna Banana Workers." *American Ethnologist* 15 (2): 328–48.

Bourgois, Philippe, and Jeff Schonberg. 2009. *Righteous Dopefiend.* Berkeley: University of California Press.

Breakthrough Institute. 2018. "A Plausible Vision to Feed the Planet." The Breakthrough Institute Working Paper, San Francisco, CA.

Brioch, M., M. Hansen, F. Stolle, P. Potapov, A. Margono, and B. Adusei. 2011. "Remotely Sensed Forest Cover Loss Shows High Spatial and Temporal Variation across Sumatra and Kalimantan, Indonesia 2000–2008." *Environmental Research Letters* 6 (1): 014010.

Brookfield, Harold. 2001. *Exploring Agrodiversity.* New York: Columbia University Press

———. 2008. "Family Farms Are Still Around: Time to Invert the Old Agrarian Question." *Geography Compass* 2 (1): 108–26.

Brookfield, Harold, and H. Parson. 2007. *Family Farms: Survival and Prospect; A World-Wide Analysis.* London: Routledge

Brosius, Peter. 1999. "Green Dots, Pink Hearts: Displacing Politics from the Malaysian Rain Forest." *American Anthropologist* 101 (1): 36–57.

Carlson, Kimberly M., D. Ratnasari, A. Pittman, B. Soares-Filho, G. Asner, S. Trigg, D. Gaveau, D. Lawrence, and H. Rodrigues. 2012. "Committed Carbon Emissions, Deforestation, and Community Land Conversion from Oil Palm Plantation Expansion in West Kalimantan, Indonesia." *Proceedings of the National Academy of Sciences of the United States of America* 109 (19): 7559–64.

Cattaneo, Claudio, and M. Gavaldà. 2010. "The Experience of Urban Squats in Collserola, Barcelona: What Kind of Degrowth?" *Journal of Cleaner Production* 18: 581–89.

Chayanov, A. V. (1922) 1986. *The Theory of Peasant Economy.* Madison: University of Wisconsin Press.

Cheung, S. 1969 *The Theory of Share Tenancy.* Chicago: University of Chicago Press.

Clastres, Pierre. 1977. *Society against the State: The Leader as Servant and Human Uses of Power among the Indians of the Americas.* New York: Urizen Press.

Conklin, Harold. 1957. *Hanunoo Agriculture: A Report on an Integral System of Shifting cultivation in the Philippines.* Rome: Food and Agriculture Organization.

Connell, Joseph H. 1978. "Diversity in Tropical Rain Forests and Coral Reefs." *Science* 199 (4335): 1302–10.

Cribb, Robert. 2002. "Unresolved Problems in the Indonesian Killings of 1965–1966." *Asian Survey* 42 (4): 550–63.

Darwin, Charles. 1837. *Notebook B: Transmutation of species* (1837–1838). Transcribed by Kees Rookmaaker. Darwin Online. http://darwin-online.org.uk/manuscripts.html.

De Moor, Martina, Paul Warde, and Leigh Shaw-Taylor. 2002. *The Management of Common Land in North West Europe, c. 1500–1850*. Turnhout: Brepols Publishers.

Dias Martins, Monica. 2006. "Learning to Participate: The MST Experience in Brazil." In *Promised Land: Competing Visions of Agrarian Reform*, edited by P. Rosset, R. Patel, and M. Courville. Oakland, CA: Food First Books.

Dobbin, Christine. 1977. "Economic Change in Minangkabau as a Factor in the Rise of the Padri Movement, 1784–1830." *Indonesia* 23 (23): 1–38.

———. 1983. *Islamic Revivalism in a Changing Peasant Economy*. Scandinavian Institute of Asian Studies. London: Curzon Press.

Dove, Michael R. 1983. "Theories of Swidden Agriculture, and the Political Economy of Ignorance." *Agroforestry systems* 1 (2): 85–99.

———. 1985. "The Agroecological Mythology of the Javanese and the Political Economy of Indonesia." *Indonesia* 39 (April): 1–36.

———. 1986. "The Practical Reason of Weeds in Indonesia: Peasant vs. State Views of Imperata and Chromolaena." *Human Ecology* 14 (2): 163–90. https://doi.org/10.1007/BF00889237.

———. 2004. "Anthropogenic Grasslands in Southeast Asia: Sociology of Knowledge and Implications for Agroforestry." *Agroforestry Systems* 61–62 (1–3): 423–35. https://doi.org/10.1023/B:AGFO.0000029013.29092.36.

———. 2008. "Perception of Volcanic Eruption as Agent of Change on Merapi Volcano, Central Java." *Journal of Volcanology and Geothermal Research* 172 (4): 329–37.

———. 2011. *The Banana Tree at the Gate: A History of Marginal Peoples and Global Markets in Borneo*. New Haven, CT: Yale University Press.

———. 2019. "Plants, Politics, and the Imagination over the Past 500 Years in the Indo-Malay Region." *Current Anthropology* 60 (S20): S309–20. https://doi.org/10.1086/702877.

Dunlap, Alexander. 2020. "Recognizing the 'De' in Degrowth: An Anarchist and Autonomist Engagement with Degrowth." The Anarchist Library. https://theanarchistlibrary.org/library/alexander-dunlap-recognizing-the-de-in-degrowth.

Durham, William. 1979. *Scarcity and Survival: The Ecological Origins of the Soccer War*. Stanford, CA: Stanford University Press.

Edelman, Marc. 1999. *Peasants against Globalization: Rural Social Movements in Costa Rica*. Stanford, CA: Stanford University Press.

Elson, R. E. 1994. "Village Java under the Forced Cultivation System 1830–1870." Southeast Asia Publications Series, no. 25. Sydney: Asian Studies Association of Australia.

Escobar, Arturo. 2018. *Designs for the Pluriverse: Radical Interdependence, Autonomy, and the Making of Worlds*. Durham, NC: Duke University Press.

Estes, Nick. 2019. *Our History Is Our Future*. New York: Verso.

Evans-Pritchard, E. 1950. "Social Anthropology: Past and Present. The Marett Lecture, 1950." *Man* 50: 118–24.

Fairhead, James, and Melissa Leach. 1996. *Misreading the African Landscape: Society and Ecology in a Forest-Savanna Mosaic.* Cambridge: Cambridge University Press.

Fasseur, C. 1992. *The Politics of Colonial Exploitation: Java, the Dutch, and the Cultivation System.* Ithaca, NY: Cornell University Southeast Asia Program

Ferguson, James. 1999. *Expectations of Modernity: Myths and Meanings of Urban Life on the Zambian Copperbelt.* Berkeley: University of California Press.

FERN. 2001. *Forests of Fear: The Abuse of Human Rights in Forest Conflicts.* Moreton-in-Marsh, UK: FERN. www.fern.org/fileadmin/uploads/fern/Documents/Forests%20of%20fear.pdf.

Fortmann, Louise. 1985. "The Tree Tenure Factor in Agroforestry with Particular Reference to Africa." *Agroforestry Systems* 2 (4): 229–51.

Fortmann, L., C. Antinori, and N. Nabane. 2010. "Fruits of Their Labors: Gender, Property Rights, and Tree Planting in Two Zimbabwe Villages." *Rural Sociology* 62 (3): 295–314.

Fox, Jefferson, Y. Fujita, D. Ngidang, N. L. Peluso, L. Potter, N. Sukantaladewi, J. Sturgeon, and D. Thomas. 2009. "Policies, Political-Economy, and Swidden in Southeast Asia." *Human Ecology* 37: 305–22.

Gadgil, M., and R. Guha. 1993. *This Fissured Land.* Berkeley: University of California Press.

Gasson, Ruth. 1986. "Part Time Farming: Strategy for Survival?" *Sociologia Ruralis* 24 (3/4): 328–64.

Geertz, Clifford. 1963. *Agricultural Involution: The Processes of Ecological Change in Indonesia.* Berkeley: University of California Press.

Gibson-Graham, J. K. 2006. *A Post Capitalist Politics.* Minneapolis: University of Minnesota Press.

Gilbert, D., and Afrizal. 2019. "The Land Exclusion Dilemma and Sumatra's Agrarian Reactionaries." *Journal of Peasant Studies* 46 (4): 681–701.

Glover, D., and Lim Teck Ghee. 1992. *Contract Farming in Southeast Asia: Three County Studies.* Kuala Lumpur: University of Malaya, Institute for Advanced Studies.

Goldman, Emma. 1911. *Anarchism and Other Essays.* New York: Mother Earth Publishing. www.marxists.org/reference/archive/goldman/works/1911/woman-suffrage.htm.

Graeber, David. 2004. *Fragments of an Anarchist Anthropology.* Chicago: Prickly Paradigm Press.

———. 2009. *Direct Action: An Ethnography.* Oakland, CA: AK Press.

Graves, Elizabeth. 1981. *The Minangkabau Response to Dutch Colonial Role in the Nineteenth Century.* Monograph Series no. 60. Ithaca, NY: Cornell University Press.

Guha, Ramachandra. 1990. *The Unquiet Woods: Ecological Change and Peasant Resistance in the Himalaya.* Berkeley: University of California Press.

Gupta, Akhil, and James Ferguson. 1997. *Anthropological Locations: Boundaries and Grounds of a Field Science.* Berkeley: University of California Press.

Guthman, Julie. 2017. "Life Itself under Contract: Rent-Seeking and Biopolitical Devolution through Partnerships in California's Strawberry Industry." *Journal of Peasant Studies* 44 (0): 100–117.

Habermas, Jurgen. 1987. *The Theory of Communicative Action*. Vol. 2 of *Lifeworld and System: A Critique of Functionalist Reason*. Boston: Beacon Press.

Hadiz, V. 2000. "Paramilitaries: Civil Society Gets Ugly." *Jakarta Post*, May 24, 2000.

Hadler, Jeffrey. 2008. *Muslims and Matriarchs: Cultural Resilience in Indonesia through Jihad and Colonialism*. Ithaca, NY: Cornell University Press.

Hardin, Rebecca. 2011. "Concessionary Politics Property, Patronage and Political Rivalry in Central African Management." *Current Anthropology* 32 (3): 298–99. https://doi.org/10.1086/658913.

Hariyadi, Bambang, and Tamara Ticktin. 2012. "From Shifting Cultivation to Cinnamon Agroforestry: Changing Agricultural Practices among the Serampas in the Kerinci Seblat National Park, Indonesia." *Human Ecology* 40 (2): 315–25.

Harley, J. B. 1989. "Deconstructing the Map." *Cartographica* 26 (2): 1–20.

Harrison, Jill. 2008. "Abandoned Bodies and Spaces of Sacrifice: Pesticide Drift Activism and the Contestation of Neoliberal Environmental Politics in California." *Geoforum* 39 (3): 1197–1214.

Harvey, D. 1993. "The Nature of Environment: Dialectics of Social and Environmental Change." *Socialist Register* 29: 1–59.

Hecht, Susanna B. 2014. "Forests Lost and Found in Tropical Latin America: The Woodland Green Revolution." *Journal of Peasant Studies* 41 (5): 877–909.

Hecht, Susanna B., and Sassan S. Saatchi. 2007. "Globalization and Forest Resurgence: Changes in Forest Cover in El Salvador." *BioScience* 57 (8): 663–72.

Hefner, Robert W. 1990. *The Political Economy of Mountain Java: An Interpretive History*. Berkeley: University of California Press.

Hobsbawm, E. J. 1952. "The Machine Breakers." *Past and Present* 1 (1): 57–70.

——— . 1959. *Primitive Rebels*. Manchester: University of Manchester Press

——— . 1974. "Peasant Land Occupations." *Past & Present* 62: 120–52.

Holmes, Seth. 2013. *Fresh Fruit, Broken Bodies*. Berkeley: University of California Press.

Human Rights Watch. 2003. *Without Remedy: Human Rights Abuse and Indonesia's Pulp and Paper Industry*. New York: Human Rights Watch International.

——— . 2016. "Indonesia: Persecution of Gafatar Religious Group." March 26. www.hrw.org/news/2016/03/29/indonesia-persecution-gafatar-religious-group.

Hunt, C., D. Gilbertson, and G. Rushworth. 2012. "A 50,000-Year Record of Late Pleistocene Tropical Vegetation and Human Impact in Lowland Borneo" *Quaternary Science Reviews* 37: 61–80.

Hunt, C., and R. Rabett. 2014. "Holocene Landscape Intervention and Plant Food Production Strategies in Island and Mainland Southeast Asia." *Journal of Archeological Science* 51: 22–33.

Indonesian Ministry of Environment. 2010. *Second National Communications Report to the UNFCCC*. Jakarta: Ministry of Environment.

Ingold, Tim. 2000. *The Perception of the Environment: Essays of Livelihood, Dwelling and Skill*. London: Routledge.

Inter-Census Agricultural Survey. 2018. *The Result of Inter-Census Agricultural Survey 2018*. Jakarta: Badan Pusat Statistik Indonesia.

Intergovernmental Panel on Climate Change. 2013. *Climate Change 2013: The Physical Basis*. Fifth Assessment Report. Cambridge: Cambridge University Press.

Jegathesan, Mythri. 2015. "Deficient Realities: Expertise and Uncertainity among Tea Plantation Workers in Sri Lanka." *Dialectical Anthropology* 39: 255–72.

———. 2019. *Tea and Solidarity: Tamil Women and Work in Postwar Sri Lanka*. Seattle: University of Washington Press.

Kahn, J. S. 1984. "Peasant Political Consciousness in West Sumatra: A Reanalysis of the Communist Uprising of 1927." SENRI Ethnological Studies, no. 13. Osaka: Osaka National Museum of Ethnology.

———. 1993. *Constituting the Minangkabau*. Oxford: Bergc.

Kartodirdjo, Sartono. 1966. *The Peasant's Revolt of Banten in 1888: Its Conditions, Course, and Sequel*. The Hague: Martinus Nijhoff.

Kautsky, Karl. (1899) 1988. *The Agrarian Question*. London: Zwan Publications.

Kay, Cristóbal. 2008. "Reflections on Latin American Rural Studies in the Neoliberal Globalization Period: A New Rurality?" *Development and Change* 39 (6): 915–43.

Kizos, Thanasis, Jose Ignacio Marin-Guirao, Maria Eleni Georgiadi, Sofia Dimoula, Evaggelos Karatsolis, Anastasios Mpartzas, Afroditi Mpelali, et al. 2011. "Survival Strategies of Farm Households and Multifunctional Farms in Greece." *Royal Geographical Society* 177 (4): 335–46.

Kloppenburg, Jack. 2010. "Impeding Dispossession, Enabling Repossession: Biological Open Source and the Recovery of Seed Sovereignty." *Journal of Agrarian Change* 10 (3): 367–88.

Kropotkin, Peter. 1902. *The Conquest of Bread*. London: Penguin Books.

Lansing, Stephen J. 1991. *Priests and Programmers: Technologies of Power in the Engineered Landscape of Bali*. Princeton, NJ: Princeton University Press.

Lenin, V. I. (1899) 2009. *Lenin: Collected Works*. Vol. 3, *The Development of Capitalism in Russia*. 5th ed. Moscow: Progress Publishers.

Li, Tania, and P. Semedi. 2021. *Plantation Life: Corporate Occupation in Indonesia's Oil Palm Zone*. Durham, NC: Duke University Press

Li, Tania M. 2001. "Masyarakat Adat, Difference, and the Limits of Recognition in Indonesia's Forest Zone." *Modern Asian Studies* 35 (3): 645–76.

———. 2007. *The Will to Improve*. Durham, NC: Duke University Press.

———. 2014. *Land's End: Capitalist Relations on an Indigenous Frontier*. Durham, NC: Duke University Press.

Little, P., and M. Watts. 1994. *Living under Contract: Contract Farming and Agrarian Transformation in Sub-Saharan Africa*. Madison: University of Wisconsin Press.

Lubis, Indra. 2013. "Testimony: Occupying Land Is Not Outmoded or Based on Old Theory or Practice." *Journal of Peasant Studies* 40 (4): 755–61.

Lucas, Anton. 1992. "Land Disputes in Indonesia: Some Current Perspectives." *Indonesia* 53: 79–92.

Lucas, Anton, and Carol Warren. 2003. "The State, the People, and Their Mediators: The Struggle over Agrarian Law Reform in Post-New Order Indonesia." *Indonesia* 76 (76): 87–126.

Majka, L .C, and T. J. Majka. 2000. "Organizing U.S. Farm Workers: A Continuous Struggle." In *Hungry for Profit: The Agribusiness Threat to Farmers, Food, and the Environment*, edited by F. Magdoff, J. B. Foster, and F. H. Buttel. New York: Monthly Review Press.

Malaka, Tan. (1948) 1991. *From Jail to Jail*. Akron: Ohio University Press.

Marcos, S. 1992. "The Southeast in Two Winds—A Storm and a Prophecy." http://flag.blackened.net/revolt/mexico/ezln/marcos_se_2_wind.html.

———. 2003. "Another Calendar: That of Resistance." http://flag.blackened.net/revolt/mexico/ezln/2003/marcos/resistance1.html.

Margono, B. A., S. Turubanova, I. Zhuravleva, P. Potapov, A. Tyukavina, A. Baccini, and M. C. Hansen. 2012. "Mapping and Monitoring Deforestation and Forest Degradation in Sumatra (Indonesia) Using Landsat Time Series Data Sets from 1990 to 2010." *Environmental Research Letters* 7 (3): 16. https://doi.org/10.1088/1748-9326/7/3/034010.

Marsden, Terry. 1990. "Towards the Political Economy of Pluriactivity." *Journal of Rural Studies* 6 (4): 375–82.

Marsden, William. 1811. *The History of Sumatra—Containing an Account of the Government, Laws, Customs and Manner of the Native Inhabitants with Description of the Natural Productions, and a Relation of the Ancient Political State of That Island*. London: J. M'Creery, Black-Horse-Court-Longman, Hurst, Rees, Orme, and Brown.

Martinez-Alier, J. 2011. "The EROI of Agriculture and Its Use by the Via Campesina." *Journal of Peasant Studies* 38 (1): 145–60.

McGuire, G., and B. Hering. 1987. "The Indonesian Army: Harbingers of Progress or Reactionary Predators?" In *Indonesian Politics: A Reader*, edited by C. Doran. North Queensland: James Cook University.

McMichael, Philip. 1997. "Rethinking Globalization: The Agrarian Question Revisited." *Review of International Political Economy* 4 (4): 630–62. https://doi.org/10.1080/09672299708565786.

Michon, Geneviève, F. Mary, and J. Bompard. 1986. "Multistoried Agroforestry Garden System in West Sumatra, Indonesia." *Agroforestry Systems* 4 (4): 315–38.

Mintz, Sydney. 2013. "Three Ancient Colonies: Caribbean Themes and Variations." Lecture at Stanford University Department of Anthropology, December 3.

Mitchell, Timothy. 2002. *Rule of Experts: Egypt, TechnoPolitics, Modernity*. Berkeley: University of California Press

Mohan, G.. and K. Stokke. 2000. "Participatory Development and Empowerment: The Dangers of Localism." *Third World Quarterly* 21 (2): 247–68.

Molino, Jean Francois, and Daniel Sabatier. 2001. "Tree Diversity in Tropical Rain Forests: A Validation of the Intermediate Disturbance Hypothesis." *Science* 294 (November): 1702–4.

Moyo, Sam, and Paris Yeros. 2005. *Reclaiming the Land: The Resurgence of Rural Movements in Africa, Asia, and Latin America.* London: Zed Books.

Netting, Robert M. 1993. *Smallholders, Householders: Farm Families and the Ecology of Intensive, Sustainable Agriculture.* Palo Alto, CA: Stanford University Press.

Nichols, Robert. 2020. *Theft Is Property! Dispossession and Critical Theory.* Durham, NC: Duke University Press.

O'Brien, Kevin. 1996. "Rightful Resistance." *World Politics* 49: 31–55.

Oki, Akira. 1977. *Social Change in the West Sumatran Village, 1908–1945.* Canberra: Australian National University.

Orlove, B. S. 1991. Mapping Reeds and Reading Maps: The Politics of Representation in Lake Titicaca. *American Ethnologist* 18(1): 3–38.

Ortner, S. B. 2014. "Dark Anthropology and Its Others: Anthropology since the Eighties." *HAU: The Journal of Ethnographic Theory* 6 (1): 47–73.

Peebles, Curtis. 1997. *The Corona Project: America's First Spy Satellites.* Annapolis: Naval Institute Press.

Peet, R., and Michael Watts. 1996. *Liberation Ecologies: Environment, Development, and Social Movements.* London: Routledge

Peluso, N. 1995. "Whose Woods Are These? Counter-Mapping Forest Territories in Kalimantan." *Antipode* 27: 383–406.

Peluso, N., and E. Harwell. 2001. "Territory, Custom, and the Cultural Politics of Ethnic War in West Kalimantan, Indonesia." In *Violent Environments*, edited by N. Peluso and M. Watts. Ithaca, NY: Cornell University Press.

Peluso, Nancy Lee. 1992. *Rich Forests, Poor People: Resource Control and Resistance in Java.* Berkeley: University of California Press.

———. 2011. "Emergent Forest and Private Land Regimes in Java." *Journal of Peasant Studies* 38 (4): 811–36.

Peluso, Nancy Lee, Suraya Afiff, and Noer Fauzi Rachman. 2008. "Claiming the Grounds for Reform: Agrarian and Environmental Movements in Indonesia." *Journal of Agrarian Change* 8 (July): 377–407.

Pelzer, K. 1978. *Planter and Peasant: Colonial Policy and the Agrarian Struggle in East Sumatra 1863–1947.* The Hague: Martinus Nijhoff.

Perfecto, Ivette, John Vandermeer, and Angus Wright. 2009. *Nature's Matrix: Linking Agriculture, Conservation and Food Sovereignty.* London: Earthscan.

Rachman, Noer Fauzi. 2002. "Democratizing Decentralization: Local Initiatives from Indonesia." Paper delivered to the International Association for the Study of Common Property 9th Biennial Conference, Victoria Falls, Zimbabwe.

———. 2011. "The Resurgence of Land Reform Policy and Agrarian Movements in Indonesia." PhD diss., University of California at Berkeley.

Rahman, Abdul. 2018. "Egalitarian Agrarian Politics and the Authoritarian Challenge." Emancipatory Rural Politics Initiative International Conference Paper no. 1. International Institute of Social Studies, The Hague, March 17.

Reid, Anthony. 1985. "From Betel-Chewing to Tobacco-Smoking in Indonesia." *Journal of Asian Studies* 44 (3): 529.

———. 2005. *An Indonesian Frontier: Acehnese and Other Histories of Sumatra.* Singapore: Singapore University Press.

Rigg, J. 2000. *More Than Soil: Rural Change in Southeast Asia.* London: Routledge.

Robinson, Geoffrey. 2018. *The Killing Season: A History of the Indonesian Massacres, 1965–66.* Princeton, NJ: Princeton University Press.

Rocheleau, D. 1995. "Maps, Numbers, Text, and Context: Mixing Methods in Feminist Political Ecology." *Professional Geographer* 47 (4): 458–66.

———. 2005. "Maps as Power Tools: Locating Communities in Space or Situating People and Ecologies in Place?" In *Communities and Conservation: Histories and Politics of Community-based Natural Resource Management,* edited by J. P. Brosius, A. Tsing, and C. Zerner. Walnut Creek, CA: AltaMira.

Rocheleau, Dianne E., and David Edmunds. 1997. "Women, Men and Trees: Gender, Power and Property in Forest and Agrarian Landscapes." *World Development* 25 (8): 1351–71.

Roosa, John. 2006. *Pretext for Murder: The September 30th Movement and Suharto's Coup d'Etat in Indonesia.* Madison: University of Wisconsin Press.

Rosset, Peter. 2006. "Alternatives: Between the State Above and the Movement Below. In *Promised Land: Competing Visions of Agrarian Reform,* edited by P. Rosset, R. Patel, and M. Courville. Oakland, CA: Food First Books.

Rosset, Peter, Braulio Machin Sosa, Adilén Maria Roque Jaime, and Dana Rocio Ávila Lozano. 2011. "The Campesino-to-Campesino Agroecology Movement of ANAP in Cuba: Social Process Methodology in the Construction of Sustainable Peasant Agriculture and Food Sovereignty." *Journal of Peasant Studies* 38 (1): 161–91.

Rosset, Peter, Raj Patel, and Michael Courville. 2006. *Promised Land: Competing Visions of Agrarian Reform.* Oakland, CA: Food First Books.

Ruzika, I. 1978. "Forest Exploitation in Indonesia: Past and Present." *Indonesia and the Malay World* 6 (16): 10.

Safransky, Sara. 2017. "Rethinking Land Struggle in the Postindustrial City." *Antipode* 49 (4): 1079–1100.

Saito, Kohei. 2022. *Marx in the Anthropocene: Towards the Idea of Degrowth Communism.* Cambridge: Cambridge University Press.

Sanday, Peggy Reeves. 2002. *Women at the Center: Life in a Modern Matriarchy.* Ithaca, NY: Cornell University Press.

Sandker, Marieke, and A. Suwarno. 2007. "Will Forests Remain in the Face of Oil Palm Expansion? Simulating Change in Malinau, Indonesia." *Ecology and Society* 12 (2): 37.

Schroeder, Richard, and Krishnawati Suryanata. 1996. "Gender and Class Power in Agroforestry Systems." In *Liberation Ecologies: Environment, Development, and Social Movements,* edited by R. Peet and M. Watts. London: Routledge.

Scott, James C. 1976. *The Moral Economy of the Peasant: Rebellion and Resistance in Southeast Asia.* New Haven, CT: Yale University Press.

———. 1985. *Weapons of the Weak: Everyday Forms of Resistance.* New Haven, CT: Yale University Press.

———. 1999. *Seeing Like a State: How Certain Schemes to Improve the Human Condition Have Failed.* New Haven, CT: Yale University Press.

———. 2009. *The Art of Not Being Governed: An Anarchist History of Upland Southeast Asia.* New Haven, CT: Yale University Press.

Shiva, Vandana. 1993. *Monocultures of the Mind: Perspectives on Biodiversity and Biotechnology.* London: Zed Books.

Sikor, T., and C. Lund. 2009. "Access and Property: A Question of Power and Authority." *Development and Change* 40 (1): 1–22.

Simpson, Audra. 2014. *Mohawk Interruptus: Political Life across the Borders of Settler States.* Durham, NC: Duke University Press.

Sirait, Martua Thomas. 2015. "Inclusion, Exclusion and Agrarian Change: Experiences of Forest Land Redistribution in Indonesia." PhD diss., International Institute of Social Studies, Erasmus University.

Socfin. 2011. *Socfin Group History: Expertise Built up over Time.* https://www.socfin .com/en/history/.

Spencer, J. E. 1966. *Shifting Cultivation in Southeastern Asia.* Berkeley: University of California Press

Stedman-Jones, G. 2016. *Karl Marx: Greatness and Illusion.* Cambridge, MA: Harvard University Press.

Steffan-Dewenter, I., M. Kessler, J. Barkmann, M. Bos, D. Buchori, S. Erasmi, H. Faust, et al. 2007. "Tradeoffs between Income, Biodiversity, and Ecosystem Functioning during Tropical Rainforest Conversion and Agroforestry Intensification." *Proceedings of the National Academy of Sciences of the United States of America* 104 (12): 4973–78.

Stoler, Ann Laura. 1985. *Capitalism and Confrontation in Sumatra's Plantation Belt, 1870–1979.* New Haven, CT: Yale University Press

———. 1986. "Plantation Politics and Protest on Sumatra's East Coast." *Journal of Peasant Studies* 13 (2): 124–43.

Sudibyo, Agus. 2001. *Politik Media dan Pertarungan Wacana.* Yogyakarta: LKIS.

Sunderlin, W., and I. Resosudarmo. 1999. "The Effect of Population and Migration on Forest Cover in Indonesia." *Journal of Environment & Development* 8 (2): 152–69.

Tarrow, Sidney. 1998. *Power in Movement.* Cambridge: Cambridge University Press.

Taussig, Michael. 2018. *Palma Africana.* Chicago: University of Chicago Press.

Thiollay, Jean-Marc. 1995. "The Role of Traditional Agroforests in the Conservation of Rain Forest Bird Diversity in Sumatra." *Conservation Biology* 9 (2): 335–53.

Thompson, E. P. (1963) 2001. *The Making of the English Working Class*. New York: Vintage Books.

———. 1975. *Whigs and Hunters: The Origins of the Black Act*. New York: Pantheon Books.

Toledo, V. M., E. Boege, and N. Barrera-Bassols. 2010. "The Biocultural Heritage of Mexico: An Overview." *Landscape* 3: 6–10.

Tsing, Anna, Heather Swanson, Elaine Gan, and Nils Bubandt. 2017. *Arts of Living on a Damaged Planet: Ghosts and Monsters of the Anthropocene*. Minneapolis: University of Minnesota Press.

Tsing, Anna Lowenhaupt. 1999. "Becoming a Tribal Elder, and Other Green Development Fantasies." In *Transforming the Indonesian Uplands*, edited by T. Li. Milton Park, Abingdon, Oxon: Routledge.

———. 2005. *Friction: An Ethnography of Global Connection*. Princeton, NJ: Princeton University Press.

———. 2015. *Mushroom at the End of the World: On the Possibility of Life in Capitalist Ruins*. Princeton, NJ: Princeton University Press.

Vaccaro, Ismael, Krista Harper, and Seth Murray. 2016. *The Anthropology of Postindustrialism: Ethnographies of Disconnection*. New York: Routledge.

Van der Kaars, S., D. Penny, J. Tibby, J. Fluin, R. A. C. Dam, and P. Suparan. 2001. "Late Quaternary Palaeoecology, Palynology and Palaeolimnology of a Tropical Lowland Swamp: Rawu Danau, West-Java, Indonesia." *Palaeogeography, Palaeoclimatology, Palaeoecology* 171 (3–4): 185–212.

Van der Ploeg, J. 2012. *The New Peasantries: Struggles for Autonomy and Sustainability in a New Era of Empire and Globalization*. London: Earthscan.

———. 2013. *Peasants and the Art of Farming*. Halifax, NS: Fernwood Publishing.

Van Goor, C. P. 1982. *Indonesia Forestry Abstracts: Dutch Literature until about 1960*. Wageningen: Center for Agricultural Publication and Documentation.

Vandergeest, Peter, and N. Peluso. 2006. "Empires of Forestry: Professional Forestry and State Power in Southeast Asia, Part 1." *Environment and History* 12: 31–64.

Villamor, Grace B., R. Pontius, and M. Noordwijk. 2013. "Agroforests' Growing Role in Reducing Carbon Losses from Jambi (Sumatra), Indonesia." *Regional Environmental Change* 14 (2): 825–34.

Virno, Paolo. 2004. *A Grammar of the Multitude: For an Analysis of Contemporary Forms of Life*. Los Angeles: Semiotext(e).

Von Benda-Beckmann, Franz, and Keebet von Benda-Beckman. 2011. "Myths and Stereotypes about Adat Law: A Reassessment of Van Vollenhoven in the Light of Current Struggles over Adat Law in Indonesia." *Bijdragen tot de Taal-, Land- en Volkenkunde* 167 (2–3): 167–95.

———. 2013. *Political and Legal Transformations of an Indonesian Polity: The Nagari from Colonization to Decentralization*. Cambridge: Cambridge University Press.

Ward, Colin. 1973. *Anarchy in Action*. London: Independent Publishers.

Warren, Carol. 2013. "Legal Certainty for Whom? Land Contestation and Value Transformations at Gili Trawangan, Lombok." In *Land for the People: The State and Agrarian Conflict in Indonesia*, edited by A. Lucas and C. Warren. Athens: Ohio University Press.

Watts, Michael J. 1983. *Silent Violence: Food, Famine and Peasantry in Northern Nigeria*. Berkeley: University of California Press.

———. 1994. "Life under Contract: Contract Farming, Agrarian Restructuring, and Flexible Accumulation." In *Living under Contract: Contract Farming and Agrarian Transformation in Sub-Saharan Africa*, edited by P. Little and M. J. Watts. Madison: University of Wisconsin Press.

Wells, Miriam. 1996. *Strawberry Fields: Politics, Class, and Work in California Agriculture*. Ithaca, NY: Cornell University Press.

White, B. 1983. "Agricultural Involution and Its Critics: Twenty Years after Clifford Geertz." Working Paper Series no. 6. International Institute of Social Studies, The Hague.

———. 1997. "Agroindustry and Contract Farmers in Upland West Java." *Journal of Peasant Studies* 24 (3): 100–136. https://doi.org/10.1080/03066159708438644

———. 2016. "Remembering the Indonesian Peasants' Front and Plantation Workers' Union (1945–1966)." *Journal of Peasant Studies* 43 (1): 1–16. https://doi.org/10.1080/03066150.2015.1101069.

Wijaya, Hanny. 2018. "Egalitarian Peasant Struggles and the Neoliberal-Authoritarian Embrace." Emancipatory Rural Politics Initiative Conference Paper no. 20. International Institute of Social Studies, The Hague, March 17.

Wijardjo, Boedhi, and Herlambang Perdana. 2001. *Reklaiming dan Kedaulatan Rakyat*. Jakarta: Yayasan Lembaga Bantuan Hukum Indonesia.

Wiley, Jay. 2008. *The Banana: Empires, Trade Wars, and Globalization*. Lincoln: University of Nebraska Press.

Winson, Anthony. 1982. "The 'Prussian Road' of Agrarian Development: A Reconsideration." *Economy and Society* 11 (4): 381–408. https://doi.org/10.1080/0308514 8200000014.

Wiradi, Gunawan. 2000. *Reforma Agraria: Perjalanan yang Belum Berakhir*. Bogor, Indonesia: Insist Press.

Wolf, Eric R. 1955. "Types of Latin American Peasantry: A Preliminary Discussion." *American Anthropologist* 57: 452–71.

———. 1957. "Closed Corporate Peasant Communities in Mesoamerica and Central Java." *Southwestern Journal of Anthropology* 13 (1): 1–18.

Wolf, Eric R., and Sidney W. Mintz. 1957. "Haciendas and Plantations in Middle America and the Antilles." *Social and Economic Studies* 6 (3): 380–412.

World Bank. 2018. World Development Indicators. https://datacatalog.worldbank .org/dataset/world-development-indicators.

Wrigley, G. 1988. *Coffee*. Tropical Agricultural Series. New York: Longman Scientific and Technical Publishing.

ILLUSTRATION CREDITS

Photos and illustrations are by the author except as credited otherwise here.

FIGURES

Figure 4 (p. 9). *Source:* Collaboration by the Sumatran artist Wiyono and the author.

Figure 5 (p. 30). *Source:* Undated photographs from an anonymous Casiavera resident.

Figure 7 (p. 60). *Source:* Central Intelligence Agency Freedom of Information Act Electronic Reading Room Archive.

Figure 8 (p. 62). *Source:* DOI/USGS/EROS Data Archive.

Figure 12 (p. 136). *Source:* Author illustrations from walks in the agroforests on the collective land with the reclaimers who cultivated them, noting cultivars' age, location, and height.

Figure 13 (p. 138). *Source:* U.S. Geological Survey EarthExplorer and NASA EOSDIS Data Archives.

Figure 14 (p. 139). *Source:* Drawings by the author from information collected by walking two 50-by-10 meter transects in these agroforests on the collective land with the reclaimers who cultivated them, noting species, location, and height.

Figure 16 (p. 146). *Sources:* Corona (DOI/USGS/EROS Data Archive), Worldview (NASA EOSDIS Data Archive), and Landsat satellite photography (U.S. Geological Survey EarthExplorer Archive), 1964–2016; Casiavera's reclaimers' oral histories 2013–16; and documents detailing land control, 1911–2013 from Wibawa's Archive and the West Sumatra Archive and Library Agency, Padang.

Figure 17 (p. 154). *Source:* Surya Wirawan.

MAPS

Map 2 (p. 41). *Source:* Topographisch Bureau, *Goenoeng Aren: Opgenomen in 1891–1894 (Sumatra; Blad 132)*, 1895, Batavia: Topographisch Bureau, KITLV maps collection, Leiden University.

INDEX

Founded in 1893,
UNIVERSITY OF CALIFORNIA PRESS
publishes bold, progressive books and journals
on topics in the arts, humanities, social sciences,
and natural sciences—with a focus on social
justice issues—that inspire thought and action
among readers worldwide.

The UC PRESS FOUNDATION
raises funds to uphold the press's vital role
as an independent, nonprofit publisher, and
receives philanthropic support from a wide
range of individuals and institutions—and from
committed readers like you. To learn more, visit
ucpress.edu/supportus.

www.ingramcontent.com/pod-product-compliance
Lightning Source LLC
Chambersburg PA
CBHW020842270326
41928CB00006B/509